The Business Analyst/ Project Manager

The Business Analyst/ Project Manager

A New Partnership for Managing Complexity and Uncertainty

ROBERT K. WYSOCKI

John Wiley & Sons, Inc.

Published by John Wiley & Sons, Inc., Hoboken, New Jersey.

Published simultaneously in Canada.

Library of Congress Cataloging-in-Publication Data

Wysocki, Robert K.
The project manager/business analyst : a new partnership for managing complexity and uncertainty / Robert K. Wysocki.
p. cm.
Includes index.
ISBN 978-0-470-76744-3 (hardback); ISBN 978-0-470-91023-8 (ebk);
ISBN 978-0-470-91024-5 (ebk); ISBN 978-0-470-91068-9 (ebk)
1. Nutrition. 2. Energy metabolism
1. Project managers. 2. Project management. 3. Business analysts. 4. Business planning. I. Title.
HD69.P75W956 2011
658.4'04–dc22

2010018600

Contents

Foreword

In the twenty-first century, we find ourselves at a crossroads in the world of business. Corporations large and small are still our best bet to produce wealth. Innovation is essential for businesses to remain competitive, and we must be able to change on a dime. Standing still means extinction. Traditional project management practices are giving way to alternative approaches to manage the complexity and the dynamic nature of today's projects and to deliver real business value early and often. As we mature our ability to select and invest in the most valuable projects, the relatively trivial enhancement and incremental improvement projects will go by the wayside, to enable us to dedicate scarce, expert resources to projects that bring about significant change in the way we do business. So business transformation projects are only going to get bigger, more complex, more difficult, and more prevalent. Our project performance track record thus far is bleak: Nearly two-thirds of IT projects fail or are significantly challenged, resulting in unsustainable financial impacts. Clearly, we must find new ways to manage large, complex IT projects. A perfect storm of events is colliding to force us to do projects differently.

Are we up to the challenge? Are we transforming our views of IT from *IT is irrelevant but a necessary (and very costly) expense* to *Our future depends on our ability to reach business/technology optimization, thus using technology as a competitive advantage*? Are we able to transform our thinking from *project management* to *project leadership*? What does project leadership look like? It is much more about collaboration than it is about management and control. It is the formation of expert teams to make critical project decisions. Trends include the emergence of new disciplines, including professional business analysis and complex project management. In addition, project management offices (PMOs), entities that have struggled for two decades to demonstrate their value, are giving way to enterprise practice centers of excellence, responsible for integrating and maturing all the disciplines that must come together to deliver complex business solutions.

Up until now, we have fielded what I will call average project teams that have failed to get successful results in over 60% of projects where a significant change to the existing business is needed. Complexity demands superior skills in many areas of expertise: strategy, opportunity analysis, feasibility analysis, architecture, development, business acumen, financial analysis, innovation, and vision. It is no longer about the project manager managing and controlling the project. It is about a team ("It takes a village") of experts working collaborative, each taking the lead when his or her expertise is vital.

In this important work, the author takes a deep dive into the subtleties of shared leadership for two of the key project leadership roles, Project Managers (PMs) and Business Analysts (BAs). The author presents a strong argument for PMs and BAs to become not just joined at the hip but sometimes to have one and the same hip. Instead of thinking of these roles as project managers and business analysts, we must now think of them as complex project managers and enterprise business analysts. Once we make that leap, it is clear that we must groom our PMs and BAs to become vital, strategic assets. Doing this requires leadership qualities resulting from a combination of professional and personal experiences, innate personality traits, and a keen understanding of and capability in the two professions. These superior skills take years to develop. Only a limited number of BAs/PMs will ever be able to integrate the BA/PM skill set to better handle the most complex projects. We need to identify high-potential candidates now and put them on a development fast track. To do so, Wysocki states that we need to determine the current profile of our PMs and BAs; understand the demand for superior skills by looking at past, current, and future projects; determine if we have a surplus or deficit; and develop a plan to bring these resources into balance.

Part of that plan needs to include our ability to harness the power of the PM/BA partnership, because there will certainly be projects that are best served by a PM and a BA working as co-leaders. Wysocki provides us with effective tools to help determine the need for these exceptional project assets, develop a "dual career path" for their development, and make the best configuration of leadership assignments based on project characteristics. The author goes so far as to propose a new professional—the PM/BA—how we might use such a professional to greatly enhance project performance, how the PM/BA position family and career path might be defined, and what a professional development program might look like.

As our journey to mature both the project management and business analysis capabilities ensues, Wysocki's in-depth discussion of the synergies between the two roles and the advantages of a combined PM/BA role makes a valuable contribution to our ability to be successful on complex projects of the twenty-first century.

Kathleen Hass, PMP
Principal Consultant
Kathleen Hass & Associates

Acknowledgments

For the past few years I have followed the project management and business analysis thought leaders' discussion about the roles, responsibilities, and relationship between the Project Manager and the Business Analyst. I could not resist adding my own two cents' worth to the discussion with a seven-article series in *BA Times* and again in *Project Times*. The response was overwhelming and clearly pointed out the need for a foundation on which to formally discuss the issues. That resulted in this book. I think we all owe a debt of gratitude to all of those who took the time to comment in writing and in conversation. I would not have been inspired to write this book without their involvement in the discussion.

List of Abbreviations

A large number of abbreviations are used in this book. Many are used so frequently that they have become standard terminology. The next list provides each abbreviation and its full meaning.

APF	Adaptive Project Framework
APM	Agile Project Management
ASD	Adaptive Software Development
ATP	Acceptance Test Producers
BA	Business analyst
BABOK	Business Analysis Body of Knowledge
BACOE	Business Analysis Center of Excellence
BACOP	Business Analysis Community of Practice
BAO	Business Analysis Office
BPM	Business Process Management
BP^4SO	Business Process, Project, Program and Portfolio Management Office
CMMI	Capability Maturity Model Integrated
DSDM	Dynamic Systems Development Method
HRIS	Human Resource Information System
HRMS	Human Resource Management System
IIBA	International Institute of Business Analysis
JAD	Joint Application Development
IRACIS	Increased Revenue Avoid Cost Improved Service
MPx	Emertxe
PDP	Professional Development Program
PM	Project manager
PMBOK	Project Management Body of Knowledge
PMI	Project Management Institute

PMLC	Project management life cycle
PMO	Project Management Office
POS	Project Overview Statement
QA	Quality Assurance
RBS	Requirements breakdown structure
RFID	Radio Frequency Identification
ROI	Return on investment
RUP	Rational Unified Process
SDLC	Systems development Life Cycle
SWOT	Strengths, Weaknesses, Opportunities, Threats
SME	Subject matter expert
TPM	Traditional Project Management
WBS	Work breakdown structure
xPM	Extreme Project Management

Introduction

The modern-day project manager (PM) and business analyst (BA) positions both emerged onto the business scene in the early days of the information age. What a challenging time to be alive and have the opportunity to be a contributing member during a period of disruptive innovations and high change brought on by computer technology. Many of us were not too sure where we were headed or how we were going to get there. We were the pioneers, and we set out courageously not knowing what we would find.

I have been a PM and BA for my entire professional career, which now spans more than 45 years. I may be the oldest active PM/BA and have always been interested in the similarities, differences, interactions, and overlap between the two positions. PM/BA is a project manager with significant business analysis skills. This definition will suffice until we define it more precisely in Chapter 3. Based on my own recollection, the business analysis discipline has its historical roots in systems analysis beginning with the first commercial applications of computers in the 1950s. As the BA professional came into being and developed, an overlap in responsibility between the BA and the PM professionals also emerged. That overlap has given rise to another set of challenges as both professional groups jockey for position and recognition in the business community. Both professions argue for their seat at the strategy table.

The purpose of this book is to make sense out of the overlap between the PM and the BA and to establish a game plan for how these two professionals should collaborate for maximum effectiveness.

My position is that the purpose of every project is to deliver maximum business value to the client for the time, money, and human resources invested. Period. I have never held that the purpose of a project was to complete the project according to specification and within the time and cost constraints. Yes, meeting a deadline and not overspending is nice, but in the final analysis it is not really that important. The movie *Titanic* is an excellent example. It was seriously over budget and behind schedule yet became one of the highest-grossing movies in film history.

Meeting time and cost constraints puts the PM's focus in entirely the wrong place. Meeting time and cost constraints has very little to do with project success. Project success is measured in terms of business value expected compared to business value delivered. Both the PM and the BA should be making every effort to maximize business value for the time and cost invested. That puts the goals of the PM and the BA in full alignment. Whenever there is a decision to be made on how to proceed with the project, the alternative that contributes most to business value would almost always be the correct choice. If you buy my fundamental premise regarding business value, you will be able to answer the many questions posed in this book regarding the roles, responsibilities, and interactions between the PM, BA, and PM/BA.

> *Meeting time and cost constraints has very little to do with project success. Project success is measured in terms of business value expected compared to business value delivered. Both the PM and the BA should be making every effort to maximize business value for the time and cost invested. That puts the goals of the PM and the BA in full alignment.*

So who are these PM/BA professionals? Are they artifacts of the project, or are they really professionals? I set about trying to answer my own question. In November 2009 I searched three popular web-based job sites using the position title "project manager/business analyst." I found nearly 2,000 entries under that or closely related titles. The companies posting these position descriptions were looking for professionals with skills and experiences in both disciplines. I'll share some examples of those listings later in this book. Most would argue that the two positions should remain separate, yet here were 2,000 such positions being advertised, and in a down economy. What would that search produce if done during better business times? I really didn't expect to get 2,000 hits from my search, but companies clearly have such needs, however poorly defined they might be. Something is going on here that has not been adequately understood, explained, or documented. I intend to make an impact or at least get the attention of those who are in a position to do something about it.

I want to be clear on two points before we proceed further. First, we have to separate the skills profile of the professional from how that professional is used on a project. Just because an individual has a skills profile that consists of both project management and business analysis (that's my

PM/BA professional) doesn't mean that he or she is assigned the dual role of both PM and BA on the same project. I will argue that having BA skills would help the PM work more effectively with the BAs assigned to the project and vice versa. As the project becomes more uncertain and complex, having that dual skill profile will help improve the likelihood of finding an acceptable solution to the agile project. The PM/BA is the professional with that dual skill profile but so are PMs and BAs acting apart. Project management and business analysis are inseparable.

Second, there will be projects led by a PM/BA who will perform the dual role. A number of variables will come into play that lead to this staffing alternative. These will be discussed throughout the book.

Just because an individual has a skills profile that consists of both project management and business analysis doesn't mean that he or she is assigned the dual role of both PM and BA on the same project.

One of my goals in this book is to add some structure and definition to these position types. The extent to which those accompanying skills are needed in a single position is a function of complexity, criticality, duration, and a host of other factors. Perhaps the more significant goal is to define the conditions under which the merged assignment should be used. I will discuss the merged PM and BA position as in my belief that using these hybrid positions will maximize business value compared to other staffing choices. And that will be my focus. This book will be the first such contribution to the published literature on the PM/BA.

The BA and PM position types are essential to the successful execution of contemporary projects and processes. Most projects need both skill sets, however they might be shared between the PM and the BA. As the current decade progresses and agile projects continue to move to the forefront, those merged skills will become more important and even essential in some situations. At the same time, the types of projects encountered today give us some clues as to how the two positions should collaborate or be merged into one position to deliver greater business value.

In the summer of 2008 I wrote a seven-article series sharing some very preliminary thoughts on this PM/BA position. It was published in *Business Analyst Times* (May 30 through July 31) and later reprinted in *Project Times.* The reaction from the readers was overwhelming. The articles got the attention of PMs and BAs, as I had hoped they would. Most were not in favor of such a professional having both PM and BA responsibilities on the same project. Based on many of the comments I received, I think many BAs thought I was recommending replacing them with PMs who knew about BA. That was not at all my objective. Several different perspectives were shared, and I was encouraged to pursue the topic. This book is the result of that effort. I hope I have given the topic a sound foundation from which further work will be done.

Historical Context

My career as a PM began in 1963 at Texas Instruments at about the time IBM announced System 360 and OS, its operating system. It was a landmark event in the history of computing. Little did I know at the time, but it was also the wake-up call that a technological revolution was about to take place. It was a revolution that we weren't ready for. If I remember correctly, I ran projects but was called a systems consultant. I don't recall anyone in my industry carrying the title PM. There was very little in the way of tools, templates, and processes to support me. The only software that I knew about was an old IBM1130 program that I think was called Project Control System. It produced Gantt charts and project schedules from the task dependency list. One of my friends in the building construction trade introduced me to it, and I thought it would be my silver bullet. I'm self-taught in project management, and for a while that program was my silver bullet but that did not last too long. Contemporary project management practices were quite primitive. The Project Management Institute (PMI) wouldn't arrive on the scene until 1969.

As for BAs, they weren't around—at least not by that name. What we did have were computer professionals who were called systems analysts. They were the mystics from the IT department who would talk to businesspersons to gather information about what they wanted, retreat back to the IT department to concoct a solution, get the businesspersons to approve their solution, and then tell the programmers what to build and then hope for the best. Requirements gathering was primitive, and few businesspersons were satisfied with the resulting solution, but they weren't involved enough to know what to do about it. Life was tough in those early days of project management and business analysis. Businesspersons were kept at arm's length and got involved only when it was time to sign a cryptic document under the threat of a project delay if they didn't sign. The relationship between businesspersons and systems analysts was strained. Neither one understood much about the others' discipline. Every one of my colleagues, including me, was looking for silver bullets, but there were no silver bullets to be found.

The 1970s brought an interesting twist. Fourth-generation languages (Focus, Ramis, Informix, Prologue, and many others) brought the power of the computer into business departments. End user computing was the watchword. Business managers and their professional staffs had a tool they could understand and could use to develop their own solutions. A significant group of quasi-developers were now added to the computer equation. They were both a blessing and a curse. On one hand, they took some of the load off the developers and built their own applications. On the other hand, they were not trained analysts and in some cases created conflict in how data were interpreted and reported. Systems integration was not part of their game plan.

The microcomputer arrived on the scene in 1981. This was definitely another disruptive innovation. Primitive word processing was the first commercial application. Spreadsheet applications soon followed, and business managers were once again empowered to do their own analyses.

Fast forward to the twenty-first century. The systems development landscape is more mature, as are the life cycles employed. We have linear, incremental, iterative, adaptive, and extreme projects, each with its own collection of systems development and project management life cycles. In less than 40 years PMI had grown to over 300,000 members worldwide. It is the de facto professional society for PMs. The International Institute of Business Analysis (IIBA), launched in 2003, was just getting started and had a membership of just over 4,000 worldwide at the time. By January 2010 IIBA had grown to a membership of over 11,000. Size differences aside, the two organizations have a lot in common and a lot to gain through collaboration and joint ventures. The definition of the PM/BA professional tops the list. I'm honored to have an opportunity to contribute my thoughts to the definition of the career and professional development of that hybrid position.

Areas of PM and BA Overlap

There is a lot of overlap with the skills and practices of the PM and the BA. Some have posited that BAs were originally carved out of PMs and then their skills profile was broadened and deepened to the BAs that we recognize today. I see it differently. Some 40 years ago, IT managers, in response to the criticisms that their systems analysts didn't really understand the business units, considered assigning systems analysts to the business units to learn about their business processes and represent their system needs. Some were reassigned, stayed, and became the BAs of today. Others remained in IT and are the systems analysts of today. However the BA arrived on the scene, there is an overlap.

Definition of a Project Manager

Jim Lewis provides this definition: A project manager is "the person who has total responsibility for ensuring that the project is completed on time, within budget, within scope, and at the desired performance level" (*The Project Manager's Desk Reference,* 2nd ed. [New York: McGraw-Hill, 2000]). All kinds of definitions have been put forth. I prefer Jim's because it is one of the few that include business value (desired performance level) as part of the definition. To limit the definition to just meeting time, cost, and scope parameters is incomplete. The most important part of this definition speaks to the PM's responsibility to deliver expected business value. The PM shares that responsibility with the BA. That is the assumption in this book.

Definition of a Business Analyst

The Business Analysis Body of Knowledge (BABOK) provides this definition: A business analyst is a person who works as "a liaison among stakeholders, in order to elicit, analyze, communicate and validate requirements for changes to business processes, policies and information systems. The business analyst understands business problems and opportunities in the context of the requirements and recommends solutions that enable the organization to achieve its goals" (International Institute of Business Analysis, *A Guide to the IIBA Business Analysis Body of Knowledge (BABOK Guide), Version 2.0* [Marietta, GA: Author, 2009]). In practice, there seems to be a fixation on requirements to the sacrifice of business process design and improvement, but otherwise I have no particular issue with this definition, and it will be the one I use going forward. In this definition, "deliver expected business value" is expressed as "enable the organization to achieve its goals." Rather than dwell on the disparate goals that the PM and the BA have, I prefer to dwell on the goal they share: the delivery of expected business value. After all, isn't that the true measure of a successful project?

There are two types of BAs that I think we need to consider as part of the management staffing decisions that have to be made: BA Generalist and BA Specialist. Neither is more important than the other, but there is a distinction we need to understand. First of all, BAs can respond to ideas only within their frame of reference. BA Generalists with broad cross-functional expertise can relegate most ideas to a home in their experience set and be able to exploit them at some later date, thus providing an avenue for delivering sustainable business value. BA Specialists, however, will not appreciate the value in some ideas because they cannot relate the idea to their experience set. Both generalists and specialists have value, depending on the nature of the project.

BA GENERALIST For the purposes of this book, BA Generalists are BAs whose expertise is in business process design and improvement at the enterprise level. They are not subject matter experts (SMEs) for any particular business unit or business process but rather are SMEs for the generic processes that define how a business process is designed, monitored, and improved. They often have training responsibilities across all business units in the enterprise. The BA Generalist is generally a higher-level position than the BA Specialist.

BA SPECIALIST BA Specialists are BAs whose focus and expertise are in a particular business unit. They are considered SMEs for one or more processes within that business unit. BA Specialists are usually assigned full time to a single business unit and have the full confidence of business unit managers to represent them and make decisions on their behalf.

Requirements Elicitation and Management

The major area of overlap between the PM and the BA Specialist is requirements elicitation and management. Both the PM and the BA face the same challenges here. Even under the best of circumstances the experts would agree that it is very difficult, if not impossible, to identify and document complete requirements as part of the scoping phase. In the agile world, that can be done only as part of the iterative build activities in an Agile Project Management (APM) methodology. That puts the PM and the BA in direct relation to one another and clouds their roles and relationships with respect to requirements elicitation and management. Underlying it all is the need for more collaborative efforts. I hope to bring some clarification to that relationship in this book.

Process Design and Improvement

The next area of overlap is process design and improvement. In the world of PMs and their management approach, this means applying some variation of the Software Engineering Institute Capability Maturity Model and continuous project management process and practice improvement. For BA Generalists or Specialists, this means running business process improvement projects, which means having effective APM methodologies. For these kinds of projects, it would seem that the BA would be better prepared to be the PM than would our typical PM. I advocate having an subject matter expert (SME) or BA Specialist share the PM responsibilities for process design and improvement projects. The staffing options are discussed in this book.

Skill and Competency Profile

The next area of overlap is the skill and competency profile of the effective PM and the effective BA. At the senior level, I believe that the two skill profiles should be virtually identical. Some have posited that we really don't need both types assigned to the same project. We could certainly debate that point of view but if current trends continue, I would argue that a single person, whatever label you choose to use, will soon be sufficient. The decision to use a person in both roles on the same project is not that obvious. In other words, the BA should have the requisite skills and competencies to be an effective PM, and the PM should have the requisite skills and competencies to be an effective BA. Such a professional will have the skills profile of both the PM and the BA. Lacking an appropriate name right now, I'll refer to this single professional as a PM/BA. The trends are clear, and the two positions are beginning to merge from a skills perspective. We are not too far away from that day.

The skill profiles of the BA and the PM at all levels of experience have considerable overlap. Even at the entry level, a BA will have some proficiency at managing projects, and every PM will have some proficiency at business analysis. The two disciplines are inseparable. The major question is: Is there a need for a BA who is an expert PM and for a PM who is an expert BA? This book will answer that question. A large body of literature discusses how the professionals who practice these separate disciplines should work collaboratively. That literature has been extensively researched in preparation for this book.

Acceptance Testing

The Project Overview Statement (POS) will have been developed collaboratively among the PM, the BA and the client. Part of that POS is the list of three or four quantitative metrics that define project success. These metrics will define the expected business value that, if attained, signifies project success. Those metrics are often related to increased revenue, cost avoidance, and improved service. The BA in collaboration with the client and the PM defines the acceptance testing that will be done to measure success attainment. The BA should maintain the acceptance testing over the life of the project and manage the actual testing at the appropriate time. The acceptance test procedure is usually a checklist of requirements that the solution must deliver. The assumption is that meeting requirements leads to project success as measured by the POS success criteria. When all items on the list are completed, the project enters the closing phase.

Context

This book is the first to take a critical look at these two professionals and compare the similarities and differences between their roles and responsibilities. As far as staffing a project with one or both of these professionals is concerned, that decision should be based on the complexity and uncertainty associated with the project, available staff skills profile, internal organizational environment, and market conditions. These are all determinants of whether separate BA and PM professionals or the merged PM/BA professional are the best choice for staffing a project. The popular opinion favors separate BA and PM professionals, but I contend that there are conditions under which that may not be the best choice. These conditions are explored in this book.

What is clear to me is that the skill sets of the senior PMs and BAs should become more similar. Project complexity and uncertainty require it. To function effectively on agile projects, they each need to have an intimate understanding of the other's profession and the ability to function in either role. Whether they will function in both roles on a given project is another question whose answer is quite complex. I will offer an answer to that question in this book.

There is a great deal of discussion on the roles and responsibilities of these two professionals. The list of questions is ever-changing, but currently includes:

- When does a project require both a BA and a PM?
- What is the skill and competency profile of a BA as compared to a PM?
- What is the boundary between BA and PM responsibilities on a project?
- What conditions and factors suggest that it make sense for one person to carry out the responsibilities of both the BA and the PM on the same project?
- Who is responsible for gathering and documenting requirements?
- What is the role of a business analyst on an agile or extreme project?
- The Janus complex: Are the BA and the PM really one and the same professional just displaying a different perspective depending on the situation?

This book will answer all of these questions.

How This Book Is Organized

I use a number of abbreviations in the book. Some may be familiar to you others will not. I've created a list of them that can be found in the front of the book for your convenience.

Chapter 1 uses the project management life cycle to introduce the areas of overlap and collaboration between the PM and the BA.

Building an infrastructure to manage the staffing situations introduced in Chapter 1 consists of two parts. The first part is to define the specific BA, PM, and PM/BA positions. That is done in Chapters 2 through 4. Chapter 2 and 3 expand a model that I introduced in my 2009 book, *Effective Project Management: Traditional, Agile, Extreme,* 5th ed. (Hoboken, NJ: John Wiley & Sons). In Chapter 4 we take a first look at the skills profiles of the PM and BA. Appendix A contains the actual proficiency data.

The second part is to define the project landscape that the BA, PM, and PM/BA must work within. That is the topic we will take up in Chapter 5. Finally, in Chapter 6 we put the two infrastructure parts together and discuss the PM/BA position family as it operates within a complex and uncertain project landscape. Chapter 7 is my recommendation for how an organization can develop these professionals. To support their efforts, there are training providers for PM skills and BA skills but not PM/BA skills. Several of these are identified in Appendix B. Appendix C contains a few courses that I believe are appropriate for the education and training of the PM/BA.

Who Should Read This Book

The book has a very broad target audience. In priority order the target markets are listed and described below.

PMs and BAs

This is not an introductory book. You cannot learn to be a PM or a BA by reading this book. If that is your objective, I am happy to refer you to *Effective Project Management,* mentioned earlier, for project management and to Barbara A. Carkenord's *Seven Steps to Mastering Business Analysis* (Fort Lauderdale, FL: J Ross Publishing, 2008) for business analysis.

This book is for practicing PMs and BAs of all experience levels who want to gain an understanding of the working relationship between the two professions. Some of you will have occupied both positions at the same time and on the same project. There were probably good reasons for that assignment. I have some advice to help you make that a successful experience the next time you are given such an assignment. There are senior-level career and professional development opportunities for those who possess the skills and competencies of both professions. This book will help you understand those opportunities and how to plan your careers accordingly.

PMO Directors

Developing an inventory of PM/BA professionals and staffing projects accordingly is challenging. This book will give the Project Management Office (PMO) Director a clear plan for building that cadre. The key will be to have a system that tracks the supply over time of PMs and BAs and the demand for their services over time to staff the projects that will be proposed to the project portfolio. Resource availability should be the foundation of every effective project portfolio management system.

The PMO has traditionally been defined around supporting the needs of the PM. Centers of Excellence and Communities of Practice are springing up as a forum for BAs. Should the PMO be repurposed to support PMs, Bas, and PM/BAs? I believe it should, and I spend some time discussing how that might happen in Chapter 7.

Business Process Managers

Business process managers are skilled BAs but need for certain basic skills and competencies in project management. They will lead projects that span business process design and continuous process improvement. For these projects, their BA skills will be challenged and their project management skills supportive of their efforts.

CIOs

As chief information officers (CIOs) review their project portfolio, they have to compare the supply of PMs, BAs and PM/BAs versus the demand for their services. Their objective is to define a portfolio that maximizes business value. It is important that there be a balance between the supply and demand of these professionals. Creating an infrastructure to facilitate achieving the balance between supply and demand is a multifaceted effort. It includes the hiring program; recruiting and managing contractors; having a Human Resource Management System (HRMS) that supports career and professional development, a training program aligned with staffing needs, a system to forecast project types and frequency, and an internal process for growing skilled and experienced PMs and BAs. No small challenge! This book will help CIOs develop a strategy to achieve that balance.

HR Managers

Just as the CIO will try to achieve a supply/demand balance, HR managers will have to recruit employees that assist in achieving that balance. They will be instrumental in defining the PM/BA position family and the associated skills and competencies across that position family. This book will give them the specifics of that position family and a robust structure on which they can build for the future.

Training Managers

A critical component of career and professional development plans is training. Allocating the training dollars to manage the gap between supply and demand for PMs and BAs is critical. This book will provide suggested training curricula to help training managers manage that gap.

Educators and Trainers

Here is where the rubber meets the road. Those who directly interact with these new professionals for their training and development must understand just who these new professionals are. This book will help them do just that by showing them how to put their training into context. Appendix C is a first attempt at defining an integrated curriculum for the PM/BA professional.

Curriculum and Training Developers

Training providers have very extensive curriculum for PMs and for BAs but very little for PM/BAs. Training program content needs to be integrated into courses specific to the needs of the PM/BA. Some example curricula are suggested. There is an annotated bibliography of training providers whose curricula spans both PM and BA. This is given in Appendix B.

Quality Assurance Managers

One of the major roles of the BA is process improvement. That begins with a definition of performance metrics to detect out of control business processes. In addition to their BA skills, they will need to understand the prevention and intervention strategies for managing distressed projects. That calls for a PM/BA with the requisite skills. The quality assurance (QA) manager is responsible for the career and professional development of these PM/BAs. This book provides the model.

How You Will Benefit from This Book

Project failure rates are too high, and I can't help but think that a lot of that failure comes from the scoping phase of the project—specifically requirements gathering and management. The PM and the BA both have a role, but their respective responsibilities and their interactions are not clear. I hope to change that with this book.

BABOK spends several pages on requirements gathering and management and I think too few pages on enterprise analysis. I think more emphasis needs to be placed on business process design, management, and improvement. This is especially important for projects of high complexity and uncertainty. These agile projects are often critical to the effective functioning of an enterprise, but an acceptable solution can come only from a collaborative effort of the BA and the PM. The goal of the project team is to find that solution. Our best efforts for these projects need the eyes of a senior BA and a senior PM and maybe even a PM/BA for effective solution discovery. And so the issue is one of defining the PM/BA position family and the associated career development of these hybrid professionals. For the purposes of this book a position family is a collection of positions related by title or skill profile. I hope to offer some guidance with this book.

CHAPTER 1

Project Manager and Business Analyst Project Life Cycle Collaboration

This chapter lays the foundation for the rest of the book by defining the boundaries and interactions between the business analyst (BA) and the project manager (PM). The roles and responsibilities of the two are not static but rather are dynamic and are a function of several variables. Those variables include the characteristics of the project, the business environment, the maturity of the client, the market, and the skills and competencies of the cadre of PMs and BAs. This chapter defines the typical relationship of the BA and PM across the project management life cycle (PMLC). It will help structure our discussion to know where the responsibilities of the BA leave off and those of the PM begin as well as those areas where the two have collaborative responsibilities. This knowledge will later prove to be a big help as we sift and sort through the many strategies for using these professionals to best advantage on all sorts of projects.

At this point there are no established rules of engagement or criteria for deciding how these two professionals should interact with one another or whether their skills should be combined into a single position that I am calling a PM/BA. Let me say at the outset that I believe there is a need for a cadre of professionals who possess both skill sets. How those professionals are assigned to projects is a totally different matter. The professional community has taken a variety of positions on that issue, and I will present those for your consideration. It would be presumptuous to assume that because professionals possess both skill sets they are assigned both roles on the same project. That is not my intention. I will, however, offer my thoughts for your consideration and reflection.

Historical Perspective

The birth of the information age in the 1950s planted the seeds that would quickly give rise to contention between the business sides and the technical sides of the enterprise. It originally arose out of:

- Ignorance of the other's domain
- Lack of a common language
- Lack of meaningful client involvement

The contention has changed over the years but persists to this day.

Ignorance of the Other's Domain

It was not too long ago that the throw-it-over-the-wall mentality was pervasive. What today is called a project was not recognized as such in a stovepipe organization. Each business function did its part and passed the effort on to the next business function in the process. No one functioned as the overseer so there was no control over the total effort. Fortunately, the situation has changed, and the business environment of today is characterized by teams and projects.

Over the past 60 years we have seen the emergence and maturation of the systems analyst from a high-tech professional housed in the information technology (IT) department to a BA housed in a business unit of the organization. The systems analyst has matured into a professional who provides a link between the business unit and the IT department. The BA is that professional and is squarely placed in the middle of that contention. As we will discover, the BA role is challenging and demanding. It is not a static role but a dynamic role that changes as the project, business environment, and market environment changes.

During that same 60 years the IT PM has matured from a technical professional to one who understands how to apply technical skills to the business side of the enterprise. In many organizations, the systems development life cycle (SDLC) substituted for a PMLC. Since there was only one SDLC at that time (waterfall model), there was only one PMLC. The IT PM is also placed squarely in the middle of that contention.

Lack of a Common Language

It was very obvious at the beginning of the computer age that the techies spoke a language totally unintelligible to the rest of the world. Other than the techies everyone else was technically challenged. People didn't want to understand. I've lost count of the number of times I have heard clients say: “Oh, that’s a technology project, and I don’t know much about technology. Just do the project and give it to me when you are done.” By some means they didn’t totally understand, they told the techie what they wanted, signed a functional requirements document that translated their wants into a language they didn’t really understand, and hoped what they got did the job. Sometimes it did; mostly it did not.

That barrier is no longer as serious a concern as it once was, but it still exists. Clients have learned a lot about technology in the past 60 years, and the conversation is somewhat more sophisticated. But the challenge of establishing a common language still exists. It is incumbent on the BA and the PM to verify that the client understands the language of the PM and the BA and that the BA and PM understand the language of the client. It is interesting to speculate just how much of the project failure rate can be ascribed to the language barrier, especially as it impacts the requirements gathering and management process.

I will share the approach I use to establish that common language in the “Project Overview Statement” section later in this chapter.

Lack of Meaningful Client Involvement

The Standish Group has been tracking the reason for project failures for several years now. Its 2010 report listed the top 10 reasons projects become challenged (shown in Table 1.1). Notice how important the role of the BA is in neutralizing so many of these reasons. For the first time client involvement (expressed as User Input) was at the top of the list. We’ve always had client involvement but until quite recently it amounted to little more

TABLE 1.1 Top 10 Reasons for Projects to Be Challenged

	PM Responsibility	BA Responsibility
Lack of User Input		X
Incomplete Requirements and Specifications		X
Changing Requirements and Specifications		X
Lack of Executive Support		X
Technology Incompetence	X	
Lack of Resources	X	
Unrealistic Expectations		X
Unclear Objectives		X
Unrealistic Time Frames		X
New Technology	X	

Source: Standish Group, CHAOS Report, 2010.

than signing an arcane functional specification document under threat of project delay if the document was not promptly signed. That characterized the relationship between the techie and the client in the 1950s and even into the 1960s. The techie's toolkit has evolved and now includes Joint Applications Design, Rapid Applications Development, prototyping, requirements gathering, use case scenarios, business process diagramming, and a host of other processes that bring the client into active involvement in the scoping phase of the project. The BA has been instrumental in facilitating that meaningful involvement and surely helped increase the likelihood of project success.

Contemporary projects are characterized by a high degree of complexity and uncertainty. All of the simple projects are done, and those that remain do not have clear or easy-to-find solutions. The agile approach to managing projects has become the dominant approach. Testimonial data that I have collected over the past 10 years from all over the world suggests that over 70% of all projects should be managed using some type of agile approach. Those projects cannot succeed without meaningful client involvement. How to attain that involvement and maintain it over the project life cycle is not an easy matter. The BA is a critical part of that effort, which extends over the entire project life cycle.

PMs, BAs, and Projects

I want you to think of the project in its most general sense. Every project, no matter how large or small, simple or complex, clearly defined or not, strategic, tactical, or operational, falls within the scope of this book. Process design and improvement efforts, which might be led by a BA, are considered a project even if they have not been recognized as such by the enterprise. Once an effort has been labeled a project, it is subject to the project management processes and practices in place in the organization. Every project has a PM. That PM might be a BA.

> *Project management is nothing more than organized common sense. If the project management process you are following asks you to do something that doesn't make sense for the project, don't do it. If the project management process doesn't allow you to take that discretionary action, change it!*

Few organizations allow the PM the latitude that I am suggesting. One of my client organizations does allow the PM that latitude. All the PM has to do is state reasons for not using or for modifying a process step. Management gives the PM the benefit of the doubt. If the project fails, the

PM may have to defend the decision to exclude or modify a process step. For my client organization, vesting the PM with that responsibility and the authority has empowered the PM and has most assuredly improved team morale.

I also want you to think of business analysis in its most general sense. The generic value chain defined by Michael Porter in *Competitive Advantage: Creating and Sustaining Superior Performance* (New York: Free Press, 1985) bounds the business analysis function. It includes the typical primary business functions and the processes that define them (inbound logistics, operations, outbound logistics, marketing , sales and service) and supporting functions and processes that define them (firm infrastructure, human resources management, technology development, and procurement). I realize that this may extend the boundaries of what many would consider the domain of the BA, but I have good reason for this definition. My reasons will become clear later on. The primary business functions will often permanently engage the services of a BA who is a subject matter expert (SME) in the processes that support the primary business processes. These BA specialists are discussed in Chapter 6. The supporting functions will often engage the services of a BA who has a more general knowledge of BA tools, templates, and processes. These are the BA generalists and are also discussed in Chapter 6.

The BA's role and responsibility with respect to these business functions and processes is comprehensive, and it is changing. BAs are responsible for design, development, performance monitoring, quality assurance, and improvement of the business processes that define their enterprise. The *Guide to the Business Analysis Body of Knowledge* (BABOK) *Version 2.0* defines business analysis as:

> *the set of tasks and techniques used to work as a liaison among stakeholders in order to understand the structure, policies, and operations of an organization, and to recommend solutions that enable the organization to achieve its goals.*

While this definition is a general statement of what business analysis encompasses, I interpret it to include the specific areas of responsibility that I have noted earlier. I point this out because I see so many organizations that don't go much beyond requirements gathering and management as the role and responsibility of the BA. For those organizations, once the requirements document has been submitted, the BA's job is finished. That is unfortunate because it severely constrains the effectiveness and contributions the BA can provide and unnecessarily limits the roles that the BABOK definition implies. This expansive role for BAs means that they will need a skill set that includes related disciplines such as "project management, software development, quality assurance, and interaction design" (BABOK, Version 2.0). For example, agility is becoming the hallmark of a successful organization. The BA's responsibility in this changing environment is to leverage new technologies to support the business units within this agile environment. This casts BAs in a strategic role rather than in just the tactical and operational roles that are common to many enterprises. So now BAs must be fluent not only in the language of the business but also in the language of technology so they can creatively and proactively apply technology to meet the needs of their clients. The agile business of the future will depend heavily on process changes to meet its needs in a constantly changing business environment. My focus in this book is on the proper integration of the BA and the PM as a function of project type, client maturity, business environment, market situation, and staff availability.

Every project requires a PM, but not all projects require a BA.

Even those efforts that are labeled projects that are fully contained in a single department, use only department staff, and have no impact on any other department must have someone in charge of the effort. No matter how primitive the person's project management skills might be, he or she is a PM even if the position title is in the BA position family. A position family is a group of positions that have a common foundation in titles, skills, or competencies. For those projects that do not require a BA, the client must provide an empowered SME to the project team and share project management responsibility with the PM. The SME is authorized to make decisions and commitments on behalf of the client. In my consulting practice, the SME is a co-PM, and every project requires co-PMs. The co-PMs are involved in the project from day one to day last. If a project does require a BA, both the PM and the BA should be involved in the project from day one to day last as co-PMs. In this approach I assume that the BA is empowered to speak on behalf of the client. Both the PM and BA have a role in defining what will be done and in validating that it has been done. That requires their involvement across the entire project life cycle. There will be some exceptions to this, where BAs on the project team are junior members of their profession and an empowered client representative or SME is responsible for the co-PM role.

Project Communications Model

Lack of clear, honest, open, and timely communications has long been the root cause of many of the reasons for project failure. Nowhere is the impact of that deficiency more obvious than in requirements gathering and management. This should not come as a surprise to anyone, but why has it been so difficult to do anything to correct the problem? After all, we talk all the time. Why can't we get it right? A process that I have developed to minimize the communications problem is called the Conditions of Satisfaction (COS). The COS directly impacts the three reasons for contention listed earlier. See the "Project Overview Statement" section for an overview of the COS.

Figure 1.1 illustrates the basic communication linkages among the PM, BA, and client for projects where there is a BA. These linkages are in place across the entire project life cycle. Note that the PM does not have a direct communications link with the client, but only through a partnering relationship with the BA in that communiqué. That places significant responsibility on the PM and the BA. They must think as one mind and speak with one voice as they represent the client. Note also that the client does not have a direct communications link back to the PM but only through the BA. That keeps it simple for the client and avoids misinterpretations. The client always initiates communication to the project through the BA. This PM-BA partnership protects the project from reinterpretations and misinterpretations as information moves along the client to BA to PM communications chain. The PM and BA partnering to communicate with the client keeps everyone on the same page and reduces the risk of miscommunications. It also helps keep the client meaningfully engaged in the project, which is critical to project success. The more complex and uncertain the project, the more this communications model becomes a critical success factor for the project. Every PM I have known doesn't like surprises and insists everyone be on the same page with respect to the project.

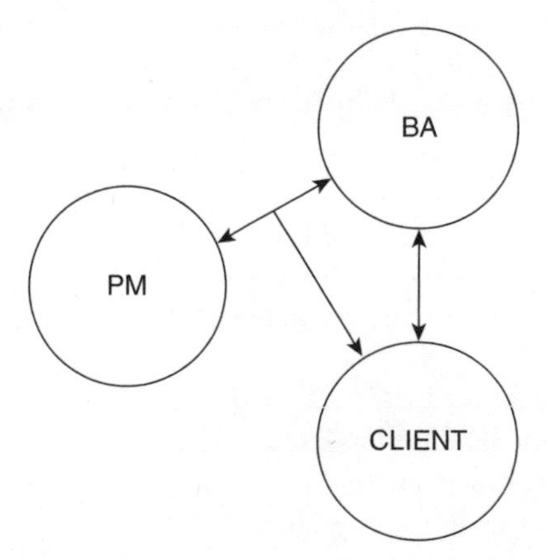

FIGURE 1.1 PM, BA, and Client Communications Mode

In those projects that do not have a BA, the communications links reduce to just one: a direct link between the client and the PM. In such situations, the PM will assume many of the responsibilities that would have been the province of the BA. This is a very simple project where the roles of PM and BA are both performed by the same professional: a PM/BA.

PM and BA Relationships across the PMLC

Table 1.2 shows a comparison over a simple linear PMLC of the responsibilities that the PM and the BA have over the life of the project. As a general observation, note that the BA has lead responsibility for the deliverables of the project and the PM has lead responsibility for managing the process that produces the deliverables. Also note that both share responsibility for project execution. The combined skills of the two positions must be sufficient to meet all project responsibilities. In this book those skills might be shared between a BA and a PM, or those skills might be possessed by a single professional with all of the project management and business analysis skills needed to execute the project effectively. There are a number of implications to this that will be discussed in detail throughout this book.

TABLE 1.2 PM and BA Responsibilities

Project Phase	Deliverable	PM	BA
Scoping	Problem Definition and Solution Validation	Support	Lead
	Project Overview Statement	Shared	Shared
	Requirements Elicitation	Support	Lead
Planning	PMLC Selection	Lead	Support
	Work Breakdown Structure	Lead	Support
	Project Plan	Lead	Support
Launching	Team Operating Rules	Shared	Shared
	Requirements Change Request	Support	Lead
	Scope Change Management	Lead	Support
	Risk Management	Shared	Shared
Monitoring and Controlling	Performance Reporting	Lead	Support
	Communications Management	Shared	Shared
Closing	Acceptance Test Procedure	Support	Lead
	Deliverables Installation	Support	Lead
	Post-Implementation Audit	Shared	Shared

In more complex projects, such as agile or extreme projects, these same five processes are present but in some form of iterative model. Planning expands to include iteration planning, which is a shared responsibility. Otherwise, the disposition of responsibilities between the PM and the BA remain the same. Agile projects can be very complex and filled with uncertainty; for that reason, they are more challenging and require a more highly skilled PM-BA partnership.

There are certain phases and parts of the PMLC that are led by either the PM or the BA, and there are certain phases and parts that are jointly led by the PM and the BA. Their performance goals are different. The PM focuses on the process, time, cost, and resource management. The BA focuses on the deliverables from the process and meeting client needs, requirements, and expected business value. These can be at odds with one another, but it is that healthy contention that produces success. The goal of their project is to find a solution that meets the expected business value that initially justified doing the project. That project goal is the driving force that helps the PM and BA resolve the contention between their performance goals.

As the PM and BA disciplines continue to mature, I'm certain we will see an integration of other professionals, such as systems developers, portfolio managers, business architects, and others. This is inevitable because to leverage technology for business value effectively, we will need to fuse the practice of these disciplines and create a collaborative project-centric environment.

Problem Definition and Solution Validation

Problem definition and solution validation is a critical deliverable because it sets the tone for the entire project. A solution cannot be defined until the problem is clearly and completely defined. The BA leads the problem definition activity. The major issue here is to drill down into the client request to define the problem. In these sessions there is always the question of wants versus needs. I have often found that what the client wants is not always what the client needs. Clients are often driven to formulate a solution to their unstated problem and offer that as what they want when in fact it is not what they need. The BA is responsible for questioning the client in an attempt to discover the real problem and hence what is needed to solve the problem. In other words, the BA has to convince clients that what they want is what they need. A root cause analysis or force field analysis can be effective aids to the BA in defining the problem.

> *What the client wants is not always what the client needs. Clients are often driven to formulate a solution to their unstated problem and offer that as what they want when in fact it is not what they need.*

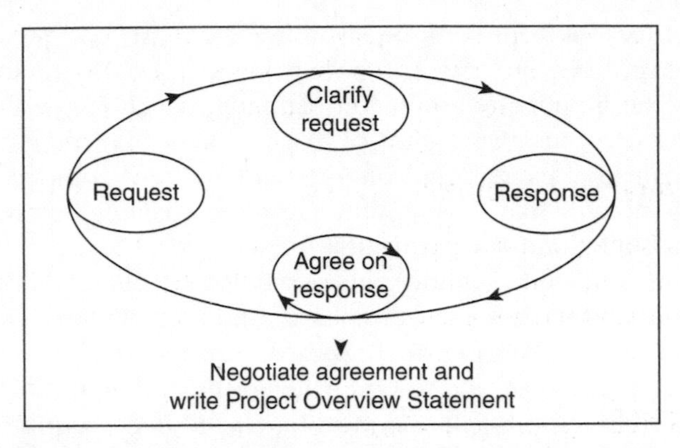

FIGURE 1.2 Conditions of Satisfaction

Armed with a definition of the problem, the BA and the PM can begin to formulate a solution or an approach to finding the solution. If an approach to finding the solution is indicated, that approach can be decided on once requirements have been documented. Chapter 5 discusses the choice of approaches in detail. Ultimately both the BA and the PM have to assure themselves that what is done in the project will align with client needs and deliver a solution that meets expected business value.

The process of defining the needs works best by using a face-to-face meeting with the client called COS. The COS establishes a common language between the PM and the client or BA. The COS process is a structured conversation between the client or BA and the PM with six steps (see Figure 1.2):

Step 1: The client and/or BA describe what they want to the PM.

Step 2: A conversation ensues in which the PM describes what she heard described. That conversation continues until the client can say to the PM: "You understand what I am asking for."

Step 3: The conversation now switches to the PM, where the PM describes what she is able to do for the client and/or BA.

Step 4: A conversation ensues in which the client describes what he heard the PM describe. That conversation continues until the PM can say to the client: "You understand what I am saying I can do for you.

Step 5: The client, the PM, and the BA understand what is wanted and what can be delivered. The two positions may not be in alignment, so a negotiation takes place to bring the request and deliverables into alignment.

Step 6: The agreement between the client and the PM is documented in the Project Overview Statement (POS) and signed by both parties.

The deliverable from the COS session is a definition of what the solution must contain in order to meet client needs. It is documented in the POS and is the input to requirements elicitation.

Project Overview Statement

The POS is a one-page document developed by the PM and the BA or the PM and client and signed by the parties who drafted it. It is the first document produced for a potential project and is documented output from a COS session.

POS CONTENTS The POS is always a one-page document written in the language of the business so that anyone who has a reason to read it will understand it. Think of the POS as a two-minute elevator speech to sell your idea to your manager. I first designed the POS based on a project request document used in the annual planning process at Texas Instruments in the 1960s. Its purpose was to organize the hundreds of requests for new products and prioritize them for investment. Executive management had to read several hundred such requests, and they didn't want to be burdened by having to read a 50-page document to make their go/no-go decisions. So the POS was strictly limited to one page. It became the filter to reduce the number of product requests and solicit more detail on those requests that passed the filter test.

There are five parts to the POS, as described next. Each part is designed to entice the executive to read on or to reject the request at that point.

1. Problem/opportunity statement
2. Project goal
3. Project objectives
4. Success criteria
5. Assumptions, risks, and obstacles

Problem/Opportunity Statement It is most important that this statement be recognized by the enterprise as a problem that has not yet been solved or an untapped business opportunity to pursue. It is a statement that does not need to be defended. The POS will be submitted to the project approval process, and you will not be there to defend this statement. Whether your project is important enough to warrant a high priority for an investment consideration is decided later. At this point the executive will have an opinion as to the importance of your problem/opportunity statement and decide whether to read your project goal statement.

Project Goal What are you proposing to do about the problem or opportunity? If it is a problem you are tackling, state here whether you are going to solve the entire problem or just some part of it. If it is an untapped business opportunity, what are you proposing to deliver through your project?

Project Objectives These are the boundary conditions for your project goal. A good format for these objectives is the SMART format (George T. Doran, "There's a SMART Way to Write Management Goals and Objectives," *Management Review* [November 1981]: 35–36):

Specific	Be specific in targeting an objective.
Measurable	Establish a measurable indicator(s) of progress.
Assignable	Make one person responsible for completion.
Realistic	State what can realistically be accomplished.
Time related	State the duration of the objective.

I like to think of each project objective as a statement of some part of the project goal. Taken together, the project objectives are a necessary and sufficient set of objectives. If all project objectives are met, the project goal is achieved. Every project objective must be accomplished in order to achieve the project goal, and no project objective is superfluous. To put it another way, think of the project goal as a pie and the project objectives as the pieces of the pie.

A project objective contains four parts: an outcome, a time frame, a quantitative measure of success, and an action. In many cases the complete statement of a project objective may be spread across several parts of the POS. This is especially true for the time frame and the success criteria.

Success Criteria The COS is the foundation of the success criteria, and the project objectives offer guidance as to what quantitative measures those criteria should include. Project success is defined as having met all of the success criteria. Some of those criteria may not be realized until long after the project has been completed. Criteria that speaks to a restoration of market share to a previous level, for example, can't be expected to happen for several quarters.

Success criteria can be of several types. You might be familiar with the acronym "IRACIS":

IR	Increase revenue
AC	Avoid cost
IS	Improve service

The important thing to keep in mind is that the success criteria must be complete. They must reflect business value that results from the project. By that I mean if the success criteria are all met, the project is deemed a success from a business value perspective. The success criteria are part of solution validation. In other words, achieving the success criteria will result in achieving the expected business value, which is often expressed as part of the goal statement.

The client and the BA are primarily responsible for establishing the success criteria. They have proposed the project on the basis that if the project is successful, it will deliver the business value that follows from achieving the success criteria. This is the return on investment (ROI) for the project. Contrary to senior management thinking, the PM is not responsible for ROI. The PM is responsible for completing successfully, the project as scoped by the client and the BA.

Assumptions, Risks, and Obstacles Certain factors will affect the outcome of the project. These factors could be:

- **Technological** (i.e., you will be using a breakthrough technology for which there is not a lot of company experience)
- **Environmental** (i.e., the market is volatile and there is a lot of new competition)
- **Interpersonal** (i.e., there is a lot of resistance to the project from some of the business units)
- **Cultural** (i.e., the project has global impact that is not well understood at this time)

These are all high-level factors and should not be confused with the more detailed risk planning yet to be done. For the purposes of the POS, you will want to call some of these high-level factors to the attention of senior management team who might have occasion to read your POS. These factors could be show stoppers, in which case the project will not get their approval. What you are hoping is that the senior managers might be able to mitigate some of these factors and improve the likelihood of project success.

PM, BA, AND CLIENT POS RESPONSIBILITIES All three parties have responsibilities with respect to the POS. The POS is the guiding document for the project team when it comes to scope discussions, so the PM must be comfortable that the POS describes a project that he or she can manage. The BA is a bridge between the PM and the client. During the COS session, the BA can help clarify understandings and should be the caretaker of the "as is" and the "to be" business process(es) that will be affected by the project. The client has to defend the business value that will result from the successful completion of the project. The client has to sell the project to management. The contents and presentation of the POS is critical to that defense, and the client should have a vested interest in its creation.

For a more detailed discussion of the COS process and its deliverable, the POS, see my book *Effective Project Management: Traditional, Agile, Extreme,* 5th ed. (Hoboken, NJ: John Wiley & Sons, 2009).

Requirements Elicitation

Requirements elicitation (aka requirements gathering) is a multiphase process that includes:

- Requirements identification
- Requirements analysis
- Requirements specification
- Requirements validation
- Requirements change management

It is not my intent to describe these processes. Refer to any good book on requirements management for details.

As the number of departments that affect or are affected by the project increases, the dynamics of the project will change. That change begins with requirements elicitation. The needs of several client departments will have to be taken into account. Here are three possible impacts that need to be considered:

1. **Scope creep during the project scoping process.** Each client department will have its list of "must haves" and "nice to haves." Not all of these will be compatible across departments, but one thing is for certain: These differences will cause scope creep. You may have to think about versioning the project—that is, decomposing it into several versions or releases.
2. **A higher incidence of "needs contention,"** which means the needs from two or more client departments may contradict one another. The BA will have to resolve the conflicts as part of validating requirements.
3. **The choice of PMLC model used to manage the project.** This decision is led by the PM. As the project becomes more of an enterprise-wide project, its likelihood of becoming a multiple team project increases. There are several implications if this should occur.

Choosing a BA to lead the requirements elicitation process is often automatic. I have never advocated the project's PM for this task. During requirements gathering, the PM should focus on the management considerations of the project as the requirements are identified and documented. If the PM is facilitating requirements gathering, he or she cannot focus on management considerations. Other than the BA or PM, someone not associated with the project could lead the requirements elicitation process. That could be an outside consultant. Having a BA associated with the project is preferred to having an outside consultant. An outside consultant does not have SME expertise. What the outside consultant brings is extensive experience facilitating requirements elicitation. The BA may not have that depth of experience. That fact should be considered in making the decision as to who will facilitate the requirements-gathering session.

If multiple business units are involved in the project or if the project involves an enterprise-wide process, the choice of requirements-gathering process and facilitator is a bit more complex. There are three approaches that I have used with success:

1. Centralized requirements elicitation
2. Shuttle diplomacy requirements elicitation
3. Outsourcing requirements elicitation

They are described next.

CENTRALIZED REQUIREMENTS ELICITATION In a sense, in this approach, you get all vested parties together at one time and duke it out. Every affected business unit must be represented by an individual vested with decision-making authority for his or her business unit. Nothing less will do! Each of those representatives might be an SME from their business unit or a BA with detailed knowledge of the business process(es) being investigated.

Centralized requirements elicitation does not scale well. I do not recommend it for enterprise-wide projects. When the size of the requirements elicitation team is greater than 10, it begins to be difficult to manage. While this approach can be much more contentious than shuttle diplomacy, it can be completed in much less total time.

SHUTTLE DIPLOMACY REQUIREMENTS ELICITATION Work with each business unit separately to gather their requirements and then resolve the between-business-unit contradictions through shuttle diplomacy. While this creates a more peaceful elicitation experience than the centralized approach and might be preferred for that reason, it does have some drawbacks that have to be considered. The burden of client satisfaction rests ultimately on the shoulders of the PM with the collaboration of the appropriate BAs. To come to closure may require the best conflict resolution skills that the PM can muster. Resolving conflicts by a consensus process is usually not a good idea. Better is to find the common requirements and build a partial solution around those. If it is a system being built, different user views can often resolve some of the simpler areas of disagreement. Experience using the partial solution may lead to resolution of requirements conflicts. Once the enterprise has some experience using the partial solution, later versions can expand the scope of the solution. This strategy needs to be clearly understood by all vested parties.

Another drawback to the shuttle diplomacy approach is the loss of cross-fertilization of ideas. In a centralized approach, an idea from one business unit may spur thinking on the part of another business unit, resulting in a synergy that benefits everyone. Good communications and collaboration between the team of BAs can overcome this drawback.

On the positive side, each requirements elicitation exercise can use a different approach to requirements gathering based on client characteristics and the nuances of their business unit.

OUTSOURCING REQUIREMENTS ELICITATION Unfortunately, politics can creep into the exercise, and that is not good. If that is a concern, a consultant with similar experiences would be my choice to facilitate requirements elicitation. The consultant can be internal or external. The internal consultant can have specific business process knowledge and an enterprise view, which may be advantageous. External consultants can push back more effectively because they do not have to carry the political baggage of their actions. These sessions can be very difficult to facilitate, and only the most experienced consultants should be used. Flexibility, adaptability, problem solving, conflict resolution, prioritization rules, and decision making are particularly valuable skills for the consultant. These projects can be complex and carry a level of uncertainty. Using a consultant will give the PM and the BA an opportunity to sit back and objectively evaluate the information being generated.

Requirements Breakdown Structure There are at least a dozen approaches you might use for requirements elicitation, and it is not my intention here to present a tutorial on their use. There are several books on requirements management. Particularly good references are

- Barbara Carkenord, *Seven Steps to Mastering Business Analysis* (Fort Lauderdale, FL: J Ross Publishing, 2008)
- Ellen Gottesdiener, *Requirements by Collaboration: Workshop for Defining Needs* (Boston: Addison-Wesley, 2002)
- Kathleen B. Hass and Rosemary Hossenlopp, *Unearthing Business Requirements: Elicitation Tools and Techniques* (Vienna, VA: Management Concepts, 2007)
- Kathleen B. Hass, Don Wessels, and Kevin Brennan, *Getting It Right: Business Requirements Analysis Tools and Techniques* (Vienna, VA: Management Concepts, 2007)
- Suzanne Robertson and James Robertson, *Mastering the Requirements Process* (Boston: Addison-Wesley, 1999)

My focus will be on the need for a more collaborative effort between the BA and the PM in the process of effectively managing those requirements throughout the entire project life cycle. I have defined the requirements breakdown structure (RBS) as an artifact in the initiation phase of the project. It is the infrastructure that supports requirements management throughout the project life cycle, the choice of a life cycle model, and the choice of best-fit project management tools, templates, and processes.

The RBS is a hierarchical description of the client's needs as expressed through the requirements document. There are at most six levels of decomposition in the RBS, as shown in Figure 1.3:

Level 1: Client statement of a requirement
Level 2: Major functions needed to meet the requirement
Level 3: Subfunctions (for larger, more complex functions)
Level 4: Processes that describe a subfunction
Level 5: Activities that describe a process
Level 6: Feature(s) of an activity

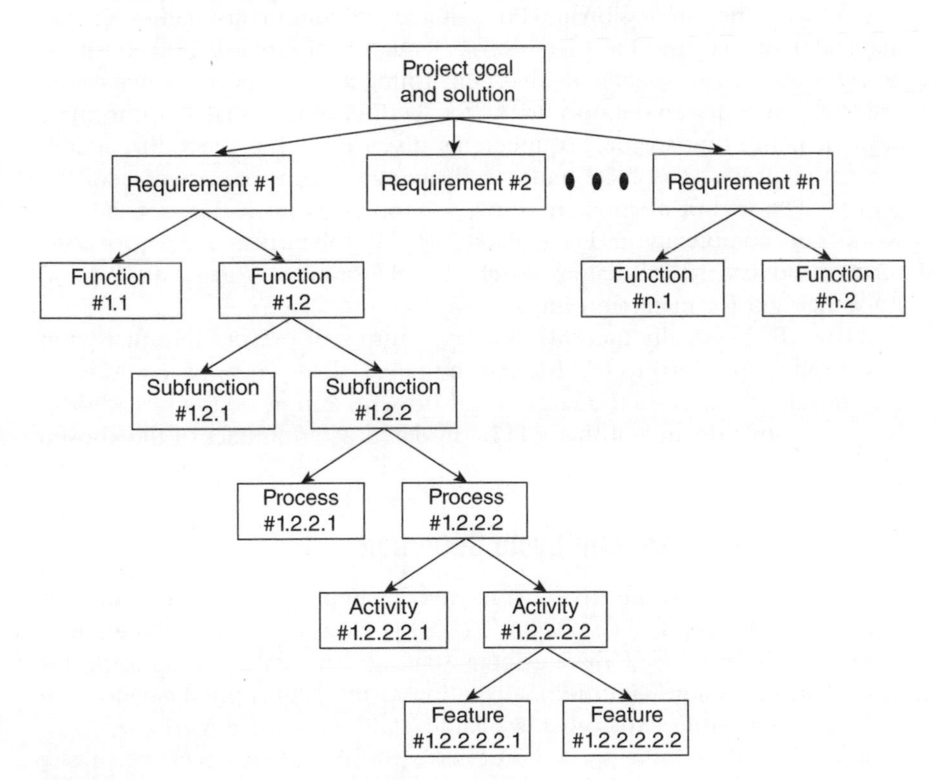

FIGURE 1.3 Requirements Breakdown Structure

The six levels are not all present in every RBS. The ones that are used are those that are needed to describe each requirement to the appropriate level of detail. So the depth of the structure is not the same for each requirement. The RBS defines what is to be done and can be thought of as the start of a deliverables-based work breakdown structure (WBS). Further decomposition of the RBS actually produces a deliverables-based WBS, which defines not only what must be done but how it will be done. There is, however, a fundamental difference between the two. The RBS may not be a complete decomposition of what will be done whereas the WBS must be complete in order for the traditional linear approaches to project management to be appropriate. There is an obvious disconnect here. The temptation is to speculate on the future to fill in the gaps in the RBS. If you take this approach, you are planting the seeds for failure.

It is this lack of completeness as portrayed in the RBS that drives the choice of PMLC and the SDLC for software development and other computer-based projects. The two life cycles are inextricably linked. Any project that produces an incomplete RBS at the outset must use some type of agile approach to managing the project. In these situations, the obvious conclusion is that the professional who manages requirements gathering and management over the life of the project must be expert at both business analysis and project management. The learning and discovery of heretofore unidentified requirements occurs in the iterations that make up an agile approach. In other words, requirements discovery takes place throughout the entire project life cycle and is fully integrated in management of the project. This is not a situation where a hand-off from a BA to a PM will work. The complexity and uncertainty of the solution and the processes for its discovery negate that approach. A collaborative effort by the PM and BA is needed for maximum impact.

The RBS is usually the PM's last opportunity to present information in a way that is intuitive to the BA and the client. After years of experimenting with the RBS, I find the hierarchical presentation to be the best choice. It is a dynamic document that will be updated as an artifact of the chosen PMLC.

Project Management Life Cycle Selection

The degree to which the RBS is considered complete will be the guide to selecting the best-fit PMLC. Chapter 5 considers this decision and the implications to the choice of management staffing. The PM is responsible for making this decision with the participation of the BA. The PM's concern is picking the best-fit PMLC that available staff and the environment can support. The BA's input to this process is the extent to which he or she can garner the requisite client involvement and what the BA needs to do to assure that involvement. So the choice of best-fit PMLC has a number of factors to consider.

Work Breakdown Structure

The project planning phase always involves the PM with active support provided by the BA and/or client representative. I don't recommend having the PM facilitate the project planning phase. I would rather see an outside consultant or some PM not related to the project do the facilitation and let the PM focus on how to manage the project based on how it is being planned.

The BA should be a member of the planning team. BAs must be empowered by the client to represent them and make decisions for them. This is critical to an effective planning session. If the planning team does not include a BA, the client must provide an SME from the business unit who has the authority to represent client interests, make decisions, and make commitments for the client during the planning sessions. In the absence of either of these situations, I have taken the hard-line approach and postponed planning until a BA or client SME representative is in place.

Creating a complete WBS for a project requires complete requirements documentation as input. For a few projects this may be the case, and a complete project plan can be created. In this situation, the support role of BAs is straightforward. They simply represent the wishes of the client as documented in the POS. Scope control during WBS creation is often a big issue. The client's appetite during planning is often bigger than the budget. You can't create a steak solution on a baloney budget. Based on client wishes, the BA can speak up whenever the WBS falls outside the scope as defined in the requirements document. The entire planning team should be watching for out of scope deliverables creeping into the WBS. When those out-of-scope deliverables are identified, they need to be discussed and the appropriate action taken. It may be that a needed deliverable was simply overlooked during requirements gathering and the project scope adjusted accordingly.

For most projects, complete requirements documentation cannot exist at the beginning of the project. In the agile project world, the complete solution is not known at the beginning of the project and hence the requirements cannot be completely known. Requirements become known only during the course of doing the project. For an agile approach, the WBS will evolve over the course of several iterations. In these situations, the role of the BA is more challenging than in the complete WBS situation. In every agile approach, the solution evolves over several iterations to find that solution, the BA must take on a more central role. My advice is to use the BA in a co-PM role. In that role, the BA is responsible for the product and the PM is responsible for the process to deliver that product. Both roles begin by defining as much of the WBS as can be generated at the beginning of the project. The WBS will be incomplete and reflect only those parts of the solution known at that time. Through BA and PM collaboration, the solution will be discovered.

Creating an initial WBS, whether it is complete or not, is difficult. It must be done right, or the project is a failure waiting to happen. If there is a BA on the team, the PM might want to let the BA facilitate the WBS creation exercise. To do that effectively, the BA will need some skills at project management. If the PM can use the BA in that role, then the PM can focus on managing the project based on how the WBS is being defined. There are many acceptable ways to build a WBS. It is not unique. The best way is the way that affords the PM the best opportunity to manage the project successfully.

Project Plan

If a complete WBS is in place, the most difficult part of project planning is already done. What remains is to estimate task duration, resource requirements, and task dependencies. Using that information, an initial project schedule can be built. It might have to be adjusted to accommodate deliverables deadlines and resource availability. The BA can be helpful in revising deliverables deadlines and release schedules based on determining requirements priorities with the client.

If the WBS is not complete, an agile approach will be used and the BA will play a more central role. First of all, he or she will be the co-PM. The added responsibility is to take the lead in updating and reprioritizing requirements in preparation for each iteration so that the PM can plan the contents of each iteration.

Team Operating Rules

When the project is approved for funding, the complete project team is recruited and comes together for their first meeting. In some settings, a core team would have been assembled to do project planning and the team membership completed only after funding approval. In any case, the first team meeting may be the first time the entire team membership will have been brought together on this project. Initially these people are not a team but just a group of people who share a common goal. They must become a team. The journey to becoming that team begins by having them decide how they are going to work together. Some processes that must be agreed to are:

- Decision making
- Problem solving
- Conflict resolution
- Team meetings

Let's take a look at each one with special reference to the PM, BA, and client interactions.

DECISION MAKING Generally speaking, it is the PM's responsibility to present the feasible alternatives that he or she can live with along with a list of the advantages and disadvantages of each as well as the schedule, cost, and resource implications. The list of alternatives may have come about through a directive, participative, or consultative model.

The situation should dictate which model makes the most sense to the PM for the specific decision situation. If alternative decisions are presented, the BA is empowered by the client to choose which of the alternatives to implement. The BA need not specify the criteria used to make that decision, but one would expect business value to be at the root of the decision.

Directive Model In the directive model, the PM makes the decisions for the entire team. While this model is expeditious, its major drawback is that the decision is based only on the information that the PM possesses. This could be complete and correct, but it might not be. A disadvantage that the PM needs to consider is that those who did not participate in the decision may be unwilling to support it. This model should be reserved for those situations where a timely decision is needed and the PM does not have the luxury of consulting others or forming a committee to define and study the alternatives. If the situation is critical to the project, however, then this model is not a good choice.

Participative Model Anyone on the team who has reason to contribute will have an opportunity to participate in the decision in the participative model. This model is a great way to empower the team and can create a synergy as the best decision is sought. Because the entire team has an opportunity to participate, the PM is more likely to gain wide support. I recommend this model whenever possible. It is politically correct too. If the project is being managed by co-PMs, they should both agree on which decision is to be taken.

Consultative Model Between the two extremes of directive and participative models is the consultative model. The PM makes the decision here, as in the directive model, but only after consulting with all affected parties or parties that may be able to contribute to the decision alternatives. Involving the entire team may be likened to killing mosquitoes with a sledgehammer. Both the directive and consultative models can be executed quickly. Politically this is a smart choice. It can gather support from the affected parties because you took the time to involve them. They will have a certain sense of empowerment too. Again, if a co-PM model is being used, both PMs should agree on the decision.

PROBLEM SOLVING The co-PMs must first define the problem and then decide who owns it. The owner will be the person charged with resolving the problem, and it is up to that person to decide how to solve it. A plan may be requested by the co-PMs. Assuming it makes business sense to solve the problem (value benefits to the project of the solution exceed the cost of implementing the solution), the co-PMs will commit the resources to support implementing the solution.

CONFLICT RESOLUTION Conflict can be good or it can be harmful. It is good when it makes the team consider alternative points of view. Here is where the contention mentioned earlier arises. The absence of this type of conflict raises the possibility of groupthink, and that can be dangerous, especially in agile projects. Conflict is harmful when it creates roadblocks and obstacles to team performance.

If the conflict is between the PM and the BA, the situation changes. They have to reach some resolution of the conflict. My recommendation is to consider the impact on the delivery of business value. That is the ultimate tiebreaker!

TEAM MEETINGS Daily team meetings are becoming the vogue for most projects. These are brief stand-up meetings attended by the PM, BA, and members of the project team who have tasks open for work at the time of the meeting. A task is open for work if its start date has passed and the task is not yet completed. The client typically does not attend. The meetings are to report project status only and not to decide where to eat lunch.

Other team meetings are problem-solving meetings that are scheduled as needed. These are attended by the problem owner and parties who are impacted by the problem and can play some role in solving it. The BA may attend these meetings in anticipation of possible impact on the client or the deliverables.

Requirements Change Request

The BA is the clearinghouse for client scope change requests. The BA and the PM will have agreed on the rules of engagement regarding the requirements change request process. I have personally experienced clients who constantly submit change requests with little regard for the impact on the project team. Many expect their requests to be honored with little or no impact on schedule, cost, or resources. The BA is an excellent buffer to protect the PM and the project team from such client behavior. The BA can provide an intake service and review and prioritize such requests before they even get to the desk of the PM. A good BA will filter these requests and submit them to the PM only if and when doing so makes sense.

Change Management

Requirements change requests are inputs to the change management process. The PM and BA are jointly responsible for defining and managing the change management process. Some of the issues to resolve include when to schedule the implementation of the change into the project plan. The BA's interest is how the release schedule will be impacted. The PM's interest will be how to accommodate the change to minimize the impact on cost, schedule, and resource requirements. The PM may present a prioritized list of alternatives that can be accommodated; the final decision for implementation is made by the BA.

Risk Management

Risk management is important across the entire PMLC. As the project becomes more complex and uncertain, risks go up significantly. The collaboration of the PM and the BA heightens with every increase in complexity or uncertainty. Chapter 6 provides the details for risk identification, risk assessment, risk response planning, and continuous monitoring of project risk.

Performance Reporting

The PM is in the best position to identify and track the metrics that will be used to periodically measure project performance. If performance falls below nominal or displays trends that if continued will seriously compromise project deliverables or lead to outright project failure, then the BA may be responsible for identifying and recommending corrective measures to the PM.

Communications Management

Developing and implementing the communications management plan is a shared responsibility between the PM and the BA. Figure 1.1 identifies the communication linkages that define the plan components. The BA-to-client linkage requires periodic reporting of project status that will be of interest to the client. These reports will be written or verbal and address scheduling problems and resolution plans, any risk status updates that affect the client, and change request status especially for open requests. The BA-to-PM linkage deals with problems and their resolution status and change request status of particular interest to the PM and the project team.

Acceptance Test Procedure

The acceptance test procedure (ATP) is led by the BA. If done properly, the ATP was initially developed by the BA with the support of the PM. The ATP is a checklist that incorporates all of the requirements with their functionality and features included as part of the checklist. As changes to the project are approved, the BA is responsible for updating the ATP so that it always accurately reflects the requirements documentation. The entire project team is responsible for conducting the ATP. The BA signs off that the deliverables have in fact met all of the specifications in the ATP.

Deliverables Installation

There are two types of deliverables (product or process), and each has its own installation process(es). Both of these installations are projects and are managed by a PM/BA. The PM provides support on an as-requested basis.

PRODUCT INSTALLATION A product installation project generally is a simple project as compared to a process installation project. A PM/BA manages this project. The PM skills of this person generally will not be as developed as those of the PM/BA who manages a process installation project.

PROCESS INSTALLATION The PM/BA who manages a process installation project should be a senior manager. This is no place for on-the-job training. Even though the ATP has been successfully completed, it was done in a test environment, which is not the same as the production environment. Now that it is time to move the process to the production environment, there will be unexpected implementation problems.

There are three installation approaches to consider and decide which to use. Each one requires its own project plan. These plans should have been completed long before installation actually occurs. Things to consider having in place before installation include user documentation, training (design, development, and delivery), and production testing.

Phased Approach In a phased approach, the process being installed will have process steps. It is these process steps that help structure the implementation sequence.

Cut-Over Approach In the cut-over approach, at some appointed time, the old process is totally replaced with the new process. This is the riskiest of the three installation approaches. More dependence is placed on the rigor of the ATP. One strategy is to have a complete test environment that is a duplicate of the production environment. An off-site backup location might be needed to support the duplicate environment and will have to be contracted.

Parallel Approach The parallel approach requires the most resources but is the least risky of the three. Any questions about the integrity and completeness of the ATP can be taken care of with this approach. Both the old process and the new process are in production status simultaneously. That allows for a comparison of old and new. The parallel approach can be used with the phased or cut-over strategy.

Post-Implementation Audit

There are two parts to the post-implementation audit. One part deals with the process that was used and how it worked or didn't work. The PM should lead that effort. The other part deals with the deliverables from the process and how well they met requirements and delivered expected business value. The BA should lead that effort.

Other Considerations and Challenges

I like to think of Table 1.2 as the standard disposition of responsibilities. A variety of factors will affect changes to this standard disposition. The roles and responsibilities of the PM and BA are not fixed. They vary from project to project, and that variance is due to the factors discussed next.

The most important consideration is that between the PM and the BA, they possess all of the skills and competencies to manage the project effectively. As part of specifying the team operating rules during the launch phase, the PM and the BA will decide what their working relationship will be and who will do what.

The most important challenge is that the division of roles and responsibilities between the PM and the BA must create an effective working relationship with the client. The purpose of the project is to maximize business value to the client and the enterprise. All division of roles and responsibilities must support that purpose. The choice of best practices through the selection of tools, templates, and processes must also support that purpose.

COMPLEXITY AND UNCERTAINTY The more complex and uncertain the project, the greater the need for both the PM and the BA to have skills and experiences in the other's discipline. They need to function more as equals on such projects. To have PMs who do not have BA skills and BAs who do not have PM skills is to add risk that flows from communication and practice variances. The objective is to have co-managers who can work together to create a synergy that increases the likelihood that they will be able to lead the project team to finding a successful solution. Complex and uncertain projects already have enough risk, and adding to it does not make good business sense. Again, having equally skilled C-Level project managers provides a backup in the event one of the members is lost to the project temporarily or permanently.

> *The more complex and uncertain the project, the more the PM and the BA will require extensive skills and experiences in the other's discipline.*

Changing technologies also add to the complexity and uncertainty of business processes. To remain competitive, organizations must be able to leverage technology for process improvement and market share protection. That requires the PM and the BA to be technology savvy.

PROJECT SIZE AND CRITICALITY A colleague of mine defines a project manager as one whose job responsibility is to manage risk. We've had many lively conversations around her perspective. Based on that definition as project size increases in terms of scope, time, cost, or resource requirements, the management structure also increases. For my comfort, that means more project management and business analysis leadership and expertise. Both the PM and the BA will need expertise in the other's discipline.

CLIENT MATURITY There must be an SME on every project. If that is not the client, then who is it? The best choice will be a BA who has the most intimate knowledge of the business function(s) involved. In the absence of a BA and the presence of a client who is technically and process challenged, the PM must compensate. That means having a carefully crafted project approach that includes workshops and other kinds of training that will keep the client meaningfully involved and able to contribute to the project. As the project becomes more complex and uncertain, the challenge to the PM is heightened.

STAFFING CHALLENGES All of these exceptions give rise to staffing challenges that can be very serious. Here is where a number of organizations fail miserably. The problem arises in those organizations that do not have a resource-constrained project portfolio management process. They will prioritize and approve projects without factoring in resource availability. It is one thing to have the needed skills among the PMs and BAs. It is another thing to have them available when and where needed.

Putting It All Together

"One size does not fit all."

That is perhaps my signature statement when it comes to project management. There are any number of factors that affect the best choice for staffing models and staffing decisions for project teams. An initial decision regarding models and decisions can even change as the project progresses and the project environment changes. Beyond staffing considerations, the best practices regarding tools, templates, and processes are subject to many of the factors discussed in this chapter. The best practice decisions will also change as the project and its environment changes. The bottom line is that project management decisions all boil down to one simple fact:

Project management is organized common sense.

CHAPTER 2

A Generic Dual-Career-Path Model

The first order of business is to establish a generic structure within which we can have an orderly and logical discussion of the BA, PM, and PM/BA professionals. The dual-career-path model presented in this chapter is sufficiently robust and does just that. For the time being I will not distinguish between the PM and BA positions. At the professional level, we will use the terms Manager and Consultant to refer to both PM and BA position levels. While I will only use three levels of manager and two levels of consultant, I want you to know that that was an arbitrary decision on my part. The discussions that follows would be the same whether there were two or three consultant position levels and two or three manager position levels. A specific company may have more or less of any of the Staff, Professional or Executive position levels. So the generic dual-career-path model introduced here is robust and fully adaptable to any organization.

Once we have an understanding of how the generic dual-career-path model works, I will embed the BA and PM positions into that structure and establish detailed position descriptions for all three levels: Staff, Professional and Executive. That is the topic of Chapter 3. Chapter 4 will establish skills profile of these position levels and types. The actual skill/proficiency data is given in Appendix A.

A Dual-Career-Path Position Family

The concept of a dual career path is not new. The British Computer Society used such a model as the infrastructure for its Professional

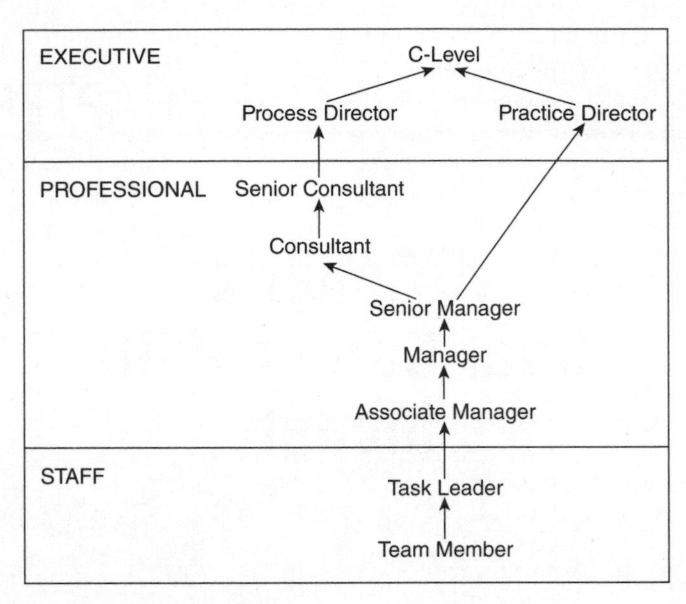

FIGURE 2.1 Generic Position Family

Development Program. The basic idea is that once people reach the professional level, they can follow a general management career path or an individual contributor (i.e., consultant) career path. Figure 2.1 displays such a model using generic position titles. Using this generic form, the model can be adapted to any position family. Originally I applied this model to a position family that integrated IT and business. For the purposes of this book, a position family is a collection of positions that are similar in terms of roles but different in terms of experience and responsibility. For example, the position family "Programmer" might have several levels, such as Coding Clerk, Junior Programmer, Programmer, and Senior Programmer. In this book I will apply it to a position family that integrates PM and BA. For specific applications additional levels of seniority can be added to fit an organization's hierarchy of positions. For example, Assistant Manager, Assistant and Associate Consultant positions can be added by an organization that has more depth in its position family.

In Figure 2.1, the leftmost path from Senior Manager through the Consultant positions to Process Director is the individual contributor path and incorporates consulting and process architecture at the professional level and process management at the executive level. At the Executive Level a Process Director owns a business process or related processes for the enterprise. So for example, the process might be the supply chain. An individual who once was a senior consultant for the supply chain now owns the supply chain processes at the director level. Other examples are the order entry/fulfillment or the project management process. In other words the consulting positions can focus on any business process including project management. For larger organizations there may be another position inserted here: Director of Process Directors, for example.

The rightmost path is the general management path. The Practice Director positions can be BA or PM based. For the PM profession an example would be Project Management Office (PMO) Director. For the BA profession the closest example would be Center of Excellence Director or Community of Practice Director. These are not formally part of the organizational chain of command and tend to provide a forum for BAs to meet and exchange practice experiences. Director-Level positions for BA have not really been implemented; at least I have not been able to find examples. Chapter 7 discusses establishing an expanded type of PMO to formally support the PM and BA profession.

At this level of abstraction, the position descriptions that follow must apply to both the PM and the BA. Having this generic description is a great aid to understanding the power in this position family. First of all, it is fully adaptable. While I only show a few levels of positions within the Staff and Professional Levels, this is not a hard-and-fast rule. Depending on the specific discipline, there could be several more levels.

This generic position family easily supports the dual career path of the PM. The PM career path begins by managing tasks, then projects, and finally PMs. First small projects (Associate PM) then larger projects (PM) and finally programs and very large complex projects (Senior PM). From the Senior PM position the individual can move to general management as a Practice Director (PMO Director, e.g.) or to a Consultant position (Risk Management Consultant, e.g.). That branch will lead to a Senior Consultant and then to Process Director (Project Management Process Architect, e.g.). Finally, the two branches merge at the Chief Information Officer (CIO) position level.

For the BA, this generic model may be ahead of its time. One of the questions I hope to offer an answer to is defining a BA career path that is more expansive than one would garner from the Business Analysis Body of Knowledge (BABOK). That is one of the reasons for my comprehensive discussion of business analysis in Chapter 1. The common point of view is that an individual enters the BA profession as a Team Member in an entry-level position and becomes a Task Leader (BA and Senior BA) after sufficient time in grade and skills are acquired and then moves into an Associate Manager position. Beginning at the Associate Manager position level, BAs are forced to follow a project management track in order to advance their careers. While that certainly is a viable career path, I think the BAs should have more career opportunities than simply project management as their ultimate career goal. Those organizations that recognize BA as a profession will define BA positions along the Associate Manager, Manager, and Senior Manager path. I will lay the foundation for this BA career path in Chapter 3 and discuss the opportunities it presents later in Chapter 7.

Staff Level

At the Staff Level, there are two positions: Team Member followed by Task Leader. This is the entry level into the PM/BA position family. Both Team Member and Task Leader positions are individual contributor positions. They are highly structured and depend on close supervision. They are not considered Professional Level positions. In a typical project both positions can report to an Associate Manager. Once staff members have acquired the experience and skills that qualifies them as professionals, they are ready for entry into the Professional Level at the Associate Manager level.

TEAM MEMBER Team Member is an entry-level position into either the project management or business analysis professions. No prior knowledge or experience in either profession is expected or required. People entering this position often come directly from completing a two- or four-year program with no full-time job experience in any business unit. At most they will have limited experience as interns or in related part-time position in project management or business analysis. In some cases they may have up to 18 months full-time experience completely outside of the project management or business analysis disciplines. An IT programmer would be a typical example.

Typical Team Member positions will be assigned to a project in a structured and supervised role. In order to be productive carrying out their first assignment, they will have to have completed initial training in using the appropriate tools, templates, and processes to be able to carry out their assigned responsibilities. Once they have acquired that working knowledge and minimal experience, they are expected to organize and plan their work to meet specified performance criteria under less direct supervision. They will quickly develop the skills to plan, schedule, and monitor their own work as well as absorb new technical information as it is presented to them.

Example position titles include Documentation Specialist, Data Analyst, and Quality Assurance (QA) Tester.

TASK LEADER Task Leader is the upper-level staff position, for people who are familiar with the scope of their tasks. After 18 to 24 months' experience as a Team Member with project management or business analysis experience, they should be qualified for promotion to Task Leader. As Task Leader they will perform tasks they are qualified for and supervise the work of Team Members assigned to their area of task responsibility.

So Task Leaders are working supervisor positions. The position is distinguished from the Team Member by the depth and complexity of their technical knowledge base and the extent to which supervision is required. Task Leaders work with little supervision and are expected to meet the requirements of their assignment under their own initiative. They will often be required to provide initial guidance and training for the less experienced Team Members assigned to their task. BAs often operate at this position level.

Task Leaders generally work unsupervised and seek advice and support only when they feel the need for such help. Their assignments are given to them with the necessary specifications for satisfactory completion, and they are expected to use the tools, templates, and processes needed for successful completion. They will have developed effective communication and problem-solving skills. They will begin to acquire skills and competencies related to their primary task assignments as a broadening experience in preparation for wider areas of responsibilities. They should begin to see the application of their tasks to broader functional areas and the business in general.

Example position titles include Business Process Analyst, Risk Analyst, Acceptance Test Lead, Requirements Elicitation Facilitator, and Change Request Intake Specialist.

Professional Level

The career path continues in the Professional Level as Associate Managers through the Senior Manager positions. As a Senior Manager the professional will make a career choice to follow an individual contributor path as a Consultant and Senior Consultant and finally as a Process Director or to a general management position at the Executive Level as a Practice Director. As an aside, I don't find any references to BAs as Consultants or Senior Consultants. However, a PMO often has senior-level PMs who function as consultants or subject matter experts (SMEs) in specialty areas of project management. It may be too early in the history of the BA to provide a career path that goes beyond the Task Leader position, but I think that artificially constrains further career growth opportunities for BAs. If business analysis is to grow and mature as a legitimate profession, these consultant positions must be part of the BA position family. I will have more to say on this later.

ASSOCIATE MANAGER Associate Manager is the entry-level position into the Professional Level. It is the lowest of three positions at the Professional Level. It normally is achieved after there is clear evidence of full competence in a specialized role. At this level, full technical accountability for work done and decisions made is expected. The ability to give technical or team leadership will have been demonstrated as well as a high degree of technical versatility and broad industry knowledge. Associate Managers often manage small projects or major parts of larger projects and are responsible to the PM. They work unsupervised and are accountable for specific results.

The typical candidate will have about 12 to 18 months of successful experience in the role of Task Leader and will have demonstrated the capacity to effectively manage simple projects and provide team leadership. Associate Managers are responsible for managing the work of staff assigned to their project but do not have direct people management responsibility.

Example position titles include Associate Project Manager and Business Analyst.

MANAGER Manager is the mid-level manager position in the Professional Level. It normally is achieved after two to four years' experience as an Associate Manager. Candidates will have demonstrated their capacity to manage projects of intermediate complexity and size. Complexity and uncertainty will characterize their assignments.

Example position titles include Project Manager, Financial Systems Business Analyst, and Business Process Improvement Analyst.

SENIOR MANAGER Senior Manager is the most senior of the three manager positions at the Professional Level. It normally is achieved after several years' experience as a Manager. Service time as a Manager is important, but more important is the candidate's demonstration of full competence in a specialized role. At this level, full technical accountability for work done and decisions made is expected. The ability to give technical or team leadership will have been demonstrated as well as a high degree of technical versatility and broad industry knowledge. The candidate will have demonstrated a capacity to manage the most complex of projects and programs and often be responsible for managing the activities of Managers and Associate Managers who function as project and subproject managers.

Example position titles include Senior Project Manager, Program Manager, Senior Business Analyst, and Business Process Owner.

CONSULTANT Consultant is the entry-level position to consulting in specialized areas of expertise across the organization. Consultants often join a project team for a limited engagement as advisors and perhaps offer training to project team members in their area of specialization. They are recognized professionals having earned that respect as Senior Managers. They work unsupervised and receive only general directions and objectives from their manager (a Process Director). If Consultants' discipline is project management, they are probably assigned to a PMO and report to the PMO Director. At the present time, a BA Consultant does not have a process home on the organizational chart. BA Centers of Excellence and Communities of Practice are coming into vogue and often offer support and advice to their more junior BAs. My recommendation is that PMOs of the future should formally support the PM/BA discipline. (See Chapter 7 for more detail.).

PMOs of the future should formally support the PM/BA discipline.

Example position titles include Enterprise Architect, Business Intelligence Consultant and Business Architect.

SENIOR CONSULTANT The Senior Consultant position differs from its more junior counterpart in that Senior Consultants are called on to advise the most senior-level executives regarding the strategic aspects of their technology specialization. This strategic advice may be shared at large project or program levels as well as at the enterprise planning levels. In either case the Senior Consultant is recognized as "the Expert" by colleagues and by executive management.

Example position titles include Senior Enterprise Architect, Business Intelligence Consultant and Senior Business Architect.

Executive Level

From Senior Manager or Senior Consultant positions, an individual can move to the Executive Level in a Practice or Process Director position. The Process Director position manages all consulting and architecture practices for the enterprise in a specific discipline. The Practice Director manages the project assignments of all Professional and even Staff positions in a specific discipline.

PROCESS DIRECTOR The Process Director position represents the level associated with the mature, relevantly experienced, and fully capable consulting professional. Such a person is fully accountable for work quality as a technical specialist. He or she possesses the background knowledge and experience to make informed and responsible decisions that are both technically sound and take the needs of the organization fully into account. Process Directors will be expected to advise and executives on strategic matters related to their technical expertise. The Director-level BA in the BA/PM or PM/BA sectors is the appropriate choice for this position.

Example position titles include VP Enterprise Architecture, VP Knowledge Management and Chief Business Architect. I would expect there to be positions called VP Business Process Management and Chief Business Architect, but I have not found any. Maybe it's too early to have such a position, but it is not too early to begin preparing candidates for such a position.

PRACTICE DIRECTOR The Practice Director is the most senior people management–level position in the PM/BA position family. It is the level occupied by the most senior manager of a business function or unit in organizations where operating effectiveness (and possibly survival) depends heavily on the function or unit and where large numbers of practitioners are deployed. A wide and deep practical knowledge base is called for, accompanied by mature management qualities.

Example position titles include PMO Director, VP of Projects, Director of IT Project Portfolio.

C-Level Position

The C-Level Position is organizationally above the PM and BA disciplines and into general executive management. CIO is a common example. A PMO Director who carries a VP designation would be another example. Whereas Director-Level positions can include operational and strategic responsibilities, the C-Level Position is strategic only.

A position titled Chief Process Officer (CPO) is beginning to appear. It is the BA counterpart of the CIO.

A position titled Chief Process Officer (CPO) is beginning to appear. It is the BA counterpart of the CIO.

Using the Dual-Career-Path Model

This model provides the structure for all PM and BA human resource utilization in the organization. A mature organization will have a comprehensive Human Resource Management System (HRMS) to support the effective development and utilization of the PM and BA resources. Having such a model is critical to the effective staffing and management of every project. Therefore it has implications to recruiting, training, professional development, project portfolio management and project staffing of both PMs and BAs.

PM Career Paths

Project management is becoming a well-established and defined profession. Table 2.1 gives examples of commonly used position titles as they distribute

TABLE 2.1 PM Position Titles across the Dual-Career-Path Model

Position Level	Position Title	PM Example
Executive	C-Level	CIO
	Process Director	PMLC Architect
	Practice Director	PMO Director
Professional	Senior Consultant/Architect	Risk Management Architect
	Consultant/Architect	Risk Management Consultant
	Senior Manager	Program Manager
	Manager	Project Manager
	Associate Manager	Testing Manager
Staff	Task Leader	Technical Lead
	Team Member	Technician

across the dual career path model. This will give you some examples to refer to as you read about more details.

BA Career Paths

The BA profession is not as well developed in terms of position titles and position descriptions. I'm going to suggest some ideas here and offer my own opinion on tentative position descriptions. Table 2.2 offers some of my thoughts. All of the example position titles are briefly discussed in the text.

TABLE 2.2 BA Position Titles across the Dual-Career-Path Model

Position Level	Position Title	BA Example
Executive	C-Level	Chief Process Officer
	Process Director	Director of Business Architecture
	Practice Director	Director BA Management
Professional	Senior Consultant/Architect	Senior BA Consultant/Architect
	Consultant/Architect	BA Consultant/Architect
	Senior Manager	Business Process Manager
	Manager	Senior BA
	Associate Manager	BA
Staff	Task Leader	BA Technician
	Team Member	Business Process Technician

BUSINESS PROCESS TECHNICIAN Those in the Business Process Technician entry-level position will work on a functional business process or other enterprise-wide support process, such as project management or procurement. Their assignments will focus on the collection, assembly, and reporting of process data. As they gain in experience, the amount of supervision they need will diminish, and they may be given a recurring assignment to complete at their own initiative. By varying their assignments, their supervisor will be preparing them for promotion into a BA Technician.

BUSINESS ANALYSIS TECHNICIAN Promotion into the Business Analysis Technician position usually follows about 18 to 24 months' experience in which people gained a working knowledge of all of the processes that support a particular business function, such as Inventory Management. Training would have included tools, templates, and processes to support the analysis of the business process to which they are assigned, such as Force Field Analysis and Root Cause Analysis. Based on process performance data gathered and reported by the Business Process Technicians they supervise, they can initiate proposals for process improvement to their manager.

BUSINESS ANALYST BAs receive and analyze performance information and process improvement needs for the processes within the scope of their responsibilities. They will propose projects to correct any anomalies they detected. They may have project management responsibilities for small projects or activities in more complex and larger improvement projects. Continuing with the Inventory Management example, they might be responsible for managing a project that examines the pick system looking for pick time reductions of 5% compared to current pick times. For an example of a larger project, they may have the same type of responsibility for reducing pick time, but now the project is looking for process improvements for the entire Inventory Management System that reduce the total cost of stocking and distributing inventory by 5%.

SENIOR BUSINESS ANALYST After having demonstrated the capacity to manage projects of intermediate complexity and size, Business Analysts are promotable to Senior BA positions. Now they are capable of managing the Inventory Management Process Improvement Project in the example just given. They can expect future assignments in larger projects of higher complexity and uncertainty.

BUSINESS PROCESS MANAGER Some organizations might use the title Business Process Owner in place of Business Process Manager. In this position, professionals will have total responsibility for the performance of the process(es) they manage. Business Process Managers work unsupervised and have demonstrated the capacity to manage or participate in large, complex projects. In this capacity they often work as full partners with a PM on projects.

BUSINESS ANALYSIS CONSULTANT/ARCHITECT The dual-career-path model branches here to either continue as an individual contributor or as a manager. BA Consultant/Architects can expect to provide consulting expertise to current BA projects for which they are recognized SMEs. This expertise could be as a subject matter expert (SME) for an enterprise-wide business process (i.e., project management life cycles [PMLSs]) or for a specific business process (Inventory Management). They can also wear the hat of Business Architect to design and build new or improved business processes.

Business Architect, at least by that name, is a new position. The best definition I have been able to find was contributed by Vadim Koteinikov:

> *Business architect is a person that initiates new business ventures or leads business innovation, designing a winning business model, and builds a sustainable balanced business system for a lasting success. (www.1000ventures.com/business_guide/business_architect.html)*

SENIOR BUSINESS ANALYSIS CONSULTANT/ARCHITECT The Senior Business Analysis Consultant/Architect position is the result of several years of broadening experiences and is given to those professionals who are recognized as enterprise-level authorities on a number of business processes. They are basically the chief architects on some part of the enterprise processes.

DIRECTOR OF BUSINESS ANALYST MANAGEMENT Directors of Business Analyst Management will be responsible for the career and professional development of all BAs and their assignment to projects. They will also inform the portfolio managers of the BA resources so that projects can be selected for the process improvement portfolio.

The Business Analyst Management Director will be the senior-most BA people management–level position in the enterprise. It closely parallels the PMO in role and responsibility.

> *The Business Analyst Management Director will be the senior-most BA people management–level position in the enterprise.*

DIRECTOR OF BUSINESS ARCHITECTURE This Executive Level position is the chief architect for the enterprise. Business Architecture Directors are responsible for the acceptable performance of all business processes.

In carrying out that responsibility, they will engage the services of the Consultant/Architects in their role as architects. Directors are strategic-level professionals as much they are operational-level professionals. They respond to senior management regarding building the process architecture needed to support the strategic goals of the organization. They also respond to operational-level performance issues by commissioning the appropriate process improvement projects.

CHIEF PROCESS OFFICER The Chief Process Officer (CPO) is equivalent to the CIO position in roles, responsibilities, and authority. Unlike the Director of Business Architecture and Director of BA Management positions, which are both strategically and operationally focused, the CPO is strategically focused.

The CPO and CIO positions are fully dependent on each other in carrying out strategic responsibilities for the enterprise. There no longer is any meaningful separation between IT and business processes. One cannot function without the other. Together these two positions function almost totally in a project-centric environment. Both depend on Agile Project Management support to carry out their responsibilities at the strategic, tactical, and operational levels.

Supply versus Demand

In order to effectively staff, manage, and complete projects, there must be a balance between the supply and demand for PMs and BAs. This balance is not a condition that exists at a point in time but rather a balance that exists across a planning horizon. So the organization should be forecasting supply over time measured against demand over time. Figure 2.2 is a graphic representation that will facilitate the next discussion.

Let me say at the outset that the optimal solution to building the supply side of the equation to satisfy the demand side of the equation does not have a closed-form mathematical solution. The best you can hope for is to utilize an HRMS that includes a decision support system that provides resource managers with the tools to view alternative staffing plans and pick what they believe to be the best alternative.

SUPPLY In order to meet the staffing requirements of the proposed project portfolio, there needs to be a balance over the planning horizon between the supply for position types and the demand for position types. The supply side shows the number of available professionals by position type over time. Some of the future supply is in the form of professionals who have development plans in place that will prepare them for assignment to a higher-level position type. They are counted in the higher-level

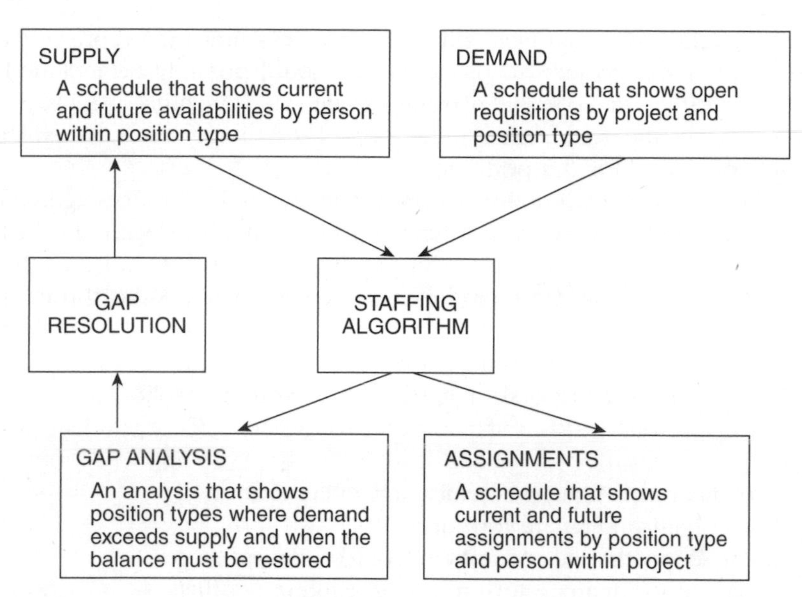

FIGURE 2.2 Supply and Demand for Staff

position beginning when they will complete their development plan. Thus, the supply profile is dynamically adjusted as development plans are complete and as project assignments are completed.

Availability can be tracked based on approved staffing assignments over time. For example, an individual who is assigned 30% to a project from July 1 through July 31 is available for a 70% assignment during that same time period. If the policy in the organization is that individuals can be assigned to only one project at a time as managers but that assignment is only 50%, they are available for the other 50% of the time, but not for assignment as project managers. Vacation time and training time are also accounted for in the schedules.

In most organizations, once the project staffing needs are identified and there is no internal candidate to fill the requirement, a position requisition is authorized and a search for an external candidate commences. As a last resort organizations turn to hiring contractors. That is the correct strategy when the skills are not needed that often and developing and maintaining that skill among your own staff does not make good business sense. In other cases, the programs to develop staff with the needed skills are in place but the numbers are not sufficient to meet project staffing needs.

DEMAND BAs are major participants in projects that are proposed for inclusion in the portfolio. In fact, submissions to the project portfolio process might best come from BAs. The business unit they represent will surely have more projects to propose than could possibly be included in the portfolio for the coming planning cycle. So BAs must prioritize the projects using the same criteria the portfolio manager uses to evaluate projects proposed for the portfolio.

I think a major contributor to project failure is having too aggressive a portfolio. Organizations do not have a good model for deciding how big is big enough when it comes to approving projects for the portfolio. I would conjecture that most organizations could increase the throughput and total business value from their portfolios by decreasing the number of projects in them. If your organization exhibits any of the following listed conditions, you need to take a hard look at your portfolio management process or put one in place if you don't have one:

- Projects are of high complexity and uncertainty.
- Project failure rates are too high.
- Client scope change requests are out of control.
- There are too many inter-project scheduling conflicts.
- There are too many eleventh-hour staffing changes.
- There are too many poor staffing assignments.
- Availability is too often treated as a skill.
- Accepting a project ignores resource availability.

Staffing Algorithm

The Staffing Algorithm can be a simple or a sophisticated tool. A simple approach involves displaying the staffing needs for each project in a prioritized list of projects. Starting at the top of the priority list, allocate staff until you come to a project that requires staff resources that are no longer available because they have been assigned to higher-priority projects.

The biggest flaw in this approach is that you will have resources that have not been assigned to a project because that project as proposed could not be fully staffed. You will have idle resources and must find things for them to do. What often happens is a negotiation between resource managers and the PMs for the unstaffed projects. PMs can make a number of adjustments that will allow their project to be staffed. These adjustments include:

- Adjusting the required skills
- Staffing a task with a less-than-qualified person and adding consulting help
- Rescheduling those tasks that cannot be staffed to times when the required skills are available
- Replanning the project

It's obvious to me that this approach can be used to improve the project portfolio and deliver more business value than the straightforward use of the simpler approach.

One of the major problems I have seen is the penchant for organizations to add projects to the portfolio when the human resources needed will not be available. If they are not available because the required skills do not exist in the staff inventory, the problem is one of recruiting or training, and the appropriate managers need to be alerted in time to do something about it. If the required skills do exist in the staff inventory, they may not be available when needed. This is a problem for the portfolio manager and is solved by:

- Not including the project in the portfolio
- Delaying the project
- Postponing projects that do not have the required skills

The PMs and BAs who have a vested interest in these projects should be involved in deciding how to go forward.

Nowhere is this problem more acute than in the assignment of PMs and BAs to projects. A model that I have used with my clients is one developed by Robert Graham and Randall Englund (*Creating an Environment for Successful Projects* [San Francisco: Jossey-Bass, 1997]). Figure 2.3 is an adaptation of their model.

As you consider the portfolio of projects, you need to take into account the ability of the PMs and BAs to deliver against that portfolio. For example,

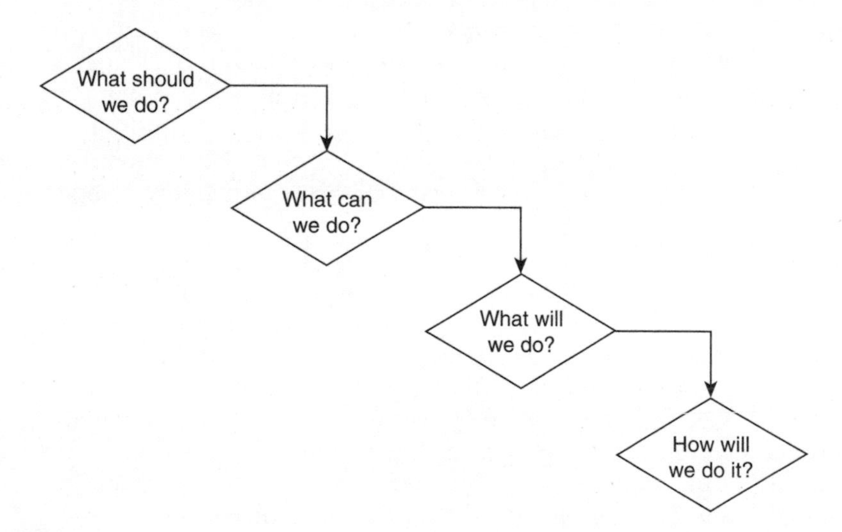

FIGURE 2.3 Graham-Englund Project Selection Model

if the portfolio were largely new or enhanced strategic applications, you would draw heavily on your most experienced and skilled BAs from the client side. If in addition those projects were characterized by a high degree of complexity and uncertainty, you would draw heavily on your senior-most PMs. What would you do with the Associate and Manager Level BAs and PMs? You want to keep them fully deployed to projects, so the portfolio should contain projects they are qualified to work on. Depending on how many Senior PMs and BAs you have available, you might want to consider someone with both BA and PM skills to manage some of these projects and function as the BA at the same time. These are very important considerations, and the Graham-Englund Project Selection Model is one model that approaches project selection with that concern in mind. Basically, it works from a prioritized list of selected projects, and staffs them until certain sets of skilled and/or experienced professionals have been fully allocated. In other words, people, not money, become the constraint on the project portfolio.

WHAT SHOULD WE DO? The answer to this question is equivalent to establishing the portfolio strategy. In the case of the Graham-Englund Project Selection Model, you are referring to the IT strategy of the organization. The answer can be found in the organization's values, mission, and objectives; it is the general direction in which the organization should be headed consistent with who it is and what it wants to be. It is IT's role to support those goals and values. IT does that by crafting a portfolio of projects consistent with those goals and values. Think of answering "What should we do?" as the demand side of the equation. You will use the project investment categories (infrastructure, maintenance, new products, and research) to identify the projects you should undertake. These categories loosely align with the skill sets of the technical staff and will give you a basis for assigning resources to projects. In fact, any categorization that allows a mapping of skills to projects will do the job. I have kept it simple for that sake of the example, but this approach can get very complex.

WHAT CAN WE DO? The answer to this question is found by comparing project requirements to the organization's resource capacity. Current commitments come into play here, as the organization must look at available capacity rather than just total capacity.

Dealing with the issue of what your organization can do raises the important issue of having a good human resource staffing model in place—one that considers future growth of the enterprise, current and projected skills inventories, training programs, career development programs, recruiting and hiring policies and plans, turnover, retirements, and so on.

TABLE 2.3 Example of Staffing Prioritized Projects at the Professional Level

Position Type	# Available	# Needed	P1	P2	P3	P4	P5	P6
Senior PM Consultant	3	2	X				X	
Senior BA Consultant	3	2	X				X	
PM Consultant	3	3		X		X		X
BA Consultant	1	2	X					X
Senior PM Manager	2	3	X		X		X	
Senior BA Manager	3	3	X		X		X	
PM Manager	3	3		X		X		X
BA Manager	3	4		X	X	X	X	
PM Associate Manager	4	4	X	X	X	X		
BA Associate Manager	4	4	X	X	X			X

Based on the data in Table 2.3, P1 through P4 can be staffed. P5 cannot be staffed because there are no Senior PM Managers or BA Managers available. There will be 2 Senior PM Consultants, 2 Senior BA Consultants, 1 PM Consultant, 1 Senior BA Manager, 1 PM Manager, and 1 BA Associate Manager unassigned. What might be done to staff P5? Here is an unexpected opportunity for off-the-job training that is often overlooked by resource managers. The hypothetical situation depicted in the table offers two opportunities for staff development that don't require sending someone out of town for training but can be done right in your own backyard. There is a need for a Senior PM Manager and a BA Manager, and there are none available; however, a PM Manager and a BA Associate Manager are available. The solution is to assign the PM Manager to P5 to fill the Senior PM Manager position request and to assign the BA Associate Manager to P5 to fill the BA Manager position request. Here you will create two opportunities for staff development, but there are some downsides to this decision. Risk might be adversely affected. Review the risk management plan for possible impact.

On the positive side, you have three choices for filling the Senior PM Manager position with a PM Manager. Pick the one who is best prepared to take on the assignment. The same is true for filling the BA Manager position with a BA Associate Manager. You have four choices. Pick the one who is best prepared to take on the assignment. You should be prepared to offer support if needed, and Consultants are available to offer that support.

WHAT WILL WE DO? The list of projects given in Table 2.3 is longer than the list of projects you will do. The creation of the "will-do" list implies that some prioritization has taken place. Various criteria, such as return on investment, break-even analysis, internal rate of return, and cost/benefit analysis, might be done to create this prioritized list.

You have assumed that a person is staffed 100% to the project. That is unlikely. In reality, a scarce resource would be scheduled to work on projects concurrently to enable more projects to be active. In practice, you would sequence the projects rather than start them all at the same time. Projects have differing durations, and this difference frees up resources to be reassigned.

HOW WILL WE DO IT? Answering this question is roughly equivalent to the selection phase in the portfolio project life cycle. In the case of resource management, "How will we do it?" is just a big staffing and scheduling problem. By scheduling scarce resources across the prioritized list, you are placing more projects on active status; that is, they will be placed in the portfolio. Detailed project plans are put in place, and the scheduling of scarce resources across the projects is coordinated. Performance against those plans is carefully monitored because the resource schedule has created a dependency between the projects. The critical chain approach to project management offers considerable detail on scheduling scarce resources across multiple projects.

Assignments

The Staffing Algorithm has two outputs. The first is a schedule that shows how available staff is assigned to projects. The second is a gap analysis, which is discussed in the next section.

Assignments are made by the Staffing Algorithm, and the availability data is updated accordingly. The sequence of those assignments being made is based on project priority.

Gap Analysis

The gap analysis can inform the HR Manager about the longer-term needs for certain positions. Through recruiting and advising, the Manager can feed the pipeline with future staff in each of the position types.

A strategy that can work is to forecast the types of projects that can be expected over time. For each of these project types, what has been the typical staffing by position type?

Gap Resolution

Resolving the gap is a matter for the career and professional development program of the enterprise. Individuals should be aware of future opportunities and encouraged to direct their professional development toward future demand in positions. The enterprise should provide added support to encourage those development efforts. It's almost like succession planning at the C Level.

Putting It All Together

We now have a generic career development model in place and ways that we can use it to staff projects with PMs and BAs. In the next chapter I will use the generic model as the infrastructure for defining the PM and BA position family in detail.

CHAPTER 3

Project Manager and Business Analyst Position Family

The details of the PM and BA position family are new with this publication. I briefly introduced the position family in my book *Effective Project Management: Traditional, Agile, Extreme*, 5th ed. (Hoboken, NJ: John Wiley & Sons, 2009). As you will discover in this chapter, the position family that I define is a very robust family that can be adapted to include all of the current and future positions in project management, business analysis, and the integration of the two. The position family that I introduce will easily adapt to any level of position detail. My expectation is that this position family will become the infrastructure for several other advances and applications in the two disciplines. It will be used in this book to:

- Define skill profiles of every position defined by the two disciplines (Chapter 4 and Appendix A).
- Support management staffing decision for projects of every type (Chapters 5 and 6).
- Establish career and professional development programs (Chapter 7).
- Recommend corporate support programs (Chapter 7).
- Advise colleges, universities, and training organizations on needed curriculum (Appendixes B and C).

The PM and BA professionals have not been shown the respect they deserve or held in high regard as a measure of what they have contributed to their organizations. Much of that baggage is due to their having inherited the sins of their grandparents. In other cases, PMs and/or BAs are not speaking from a position of strength. They tend to be too defensive when challenged. I've heard business unit managers commenting: "Now that I've got a BA, I don't need PMs anymore." Clearly there is a misunderstanding among many managers of the roles and responsibilities of PMs vis-à-vis BAs and their points of collaboration and separation. This book will help clarify those misconceptions. In writing this book, I have tried to bring together as much of the writings, presentations, and opinions on the topic of the PM and BA and their interaction on projects as are in the public domain.

Too many organizations place the responsibility for return on investment (ROI) on the shoulders of the PM and even the Project Management Office (PMO). That is totally misplaced. The PM and the PMO are responsible for establishing and providing the organization with the needed tools, templates, processes, and support, but it is up to the client and his or her BA to propose projects and validate them on the basis of the value they will return to their business unit and the organization. The client, perhaps through the BA, is responsible for ROI, and that is the position I take in this book. That ROI could be measured in terms of increased revenue, cost avoidance, and/or improved services.

Through this book I would like to help correct the misconceptions and enhance the images and credibility of the PM and BA and provide a path forward to contributing to their organizations. When the actual contributions of PMs and BAs to their organizations are understood, they should gain the respect they rightly deserve. Perhaps this book can contribute to that. In most organizations PMs and BAs are thought of as tactical and operational-level personnel when in fact they are justified in occupying a seat at the strategy table. The justification rests on the maturity and effectiveness of the project portfolio management process of the organization. The directors of PM and BA can make significant contributions to improving the performance of the portfolio. The director of project management (PMO director or VP of Projects) can recommend staffing development and assignments to improve the business value that is returned to the organization. The chief process officer seems to fit the void at the BA executive level and can prioritize projects to increase the value of BA efforts to the organization. Unfortunately, I don't know of any organization that has such an executive-level position at this time. After you read this chapter and incorporate some suggested career paths into your thinking, I hope you agree with me on the need for these executive-level positions. Until the publication of this book, these and other career paths have largely been unexplored and undefined. Not having a defined and accepted career path has probably contributed to the plight of BAs.

Chapter 2 defined a generic dual-path career model. In this chapter I will present a more detailed specification for each of these position types as they relate to the PM and the BA professional. This chapter and the next peel back layers of the onion and define the details of each PM and BA professional position type following the model illustrated in Figure 2.1. The roles and responsibilities of the PM and BA require specific skills and proficiency profiles. These vary as the level of the position increases from Staff, to Professional, to Executive and the mix of PM and BA position content varies. The skill profiles of the PM and BA positions are discussed in Chapter 4 and presented in Appendix A. The integration of the roles and responsibilities of the PM and the BA into a single position creates a family of PM/BA positions. The skill profiles of these blended positions are also discussed in Chapter 4 and presented in Appendix A.

I want to reiterate the distinction between skills profile and assigned role. I believe there is a need for professionals whose skill profile includes that of a PM and that of a BA. If the PM skill profile includes business analysis, these people will be able to work more effectively with a BA because they understand how a BA works and what he or she can be expected to do. Similarly, if the BA skill profile includes project management, these people will be able to work more effectively with a PM because they understand how a PM works and what he or she can be expected to do. In most cases this professional's role on a project will be that of either a PM or a BA. But I also believe that there are situations where this professional will be responsible for both the project management and business analysis processes on the same project. Chapter 6 discusses the conditions under which these dual role assignments arise, suggested variations to these assignments, and why they are justified.

PM and BA Position Landscape

First some background. In the early 1980s I had the opportunity to consult with the British Computer Society on the design and development of its Professional Development Program. Its program was one of the first well-defined cases of a dual-career-path model. In its case the duality came from a choice of following a management track or a consultant track as the person's career developed. In the early 1990s I had the opportunity to develop an Internet-based decision support system for IT career development for one of my clients. That system was called CareerAgent. CareerAgent was a thin client decision support system that helped IT professionals assess their skills and proficiencies, define their career goals, calculate the skills gap between their current skills profile and the profile required to achieve their career goal position, and develop a career and professional development plan to achieve their career goal. The infrastructure supporting CareerAgent was based on an integrated position family that combined IT and business functions. It paralleled the PM/BA structure used in this book.

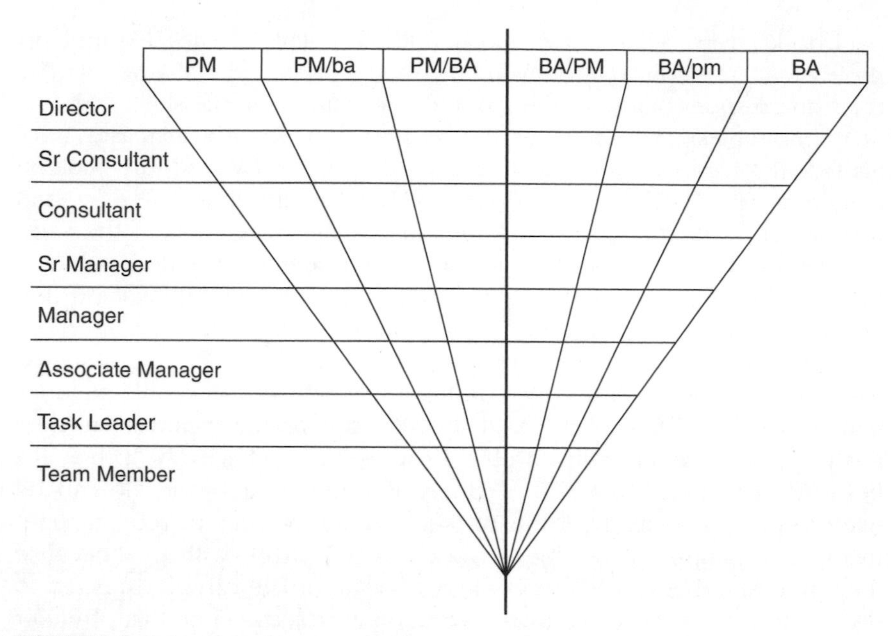

FIGURE 3.1 PM and BA Position Landscape

Much of the work I did for those two clients influenced the landscape presented in Figure 3.1.

As you probably noticed, Figure 3.1 introduced a notational change. So far I have only used PM/BA to represent a position that is some combination of PM and BA content. It is now time to change that. The notation XX/YY will now be used to denote positions whose primary discipline and position orientation is XX and whose secondary discipline is YY. Upper- and lower-case formats will also be used. For example, a position in the PM/ba sector is primarily a PM but has some ba content in the position description. Lower-case ba indicates a position that has minimal content in BA. Upper-case BA indicates a high level of content in BA. A position in the PM/BA sector is both a PM position and a BA position; staff members who occupy these positions are equally comfortable in either assignment but their position description is PM oriented. The same interpretation holds for the right half of the sectors. A position in the BA/PM sector equally represents both disciplines but the primary focus of the position is BA. At a specific level, the content of a PM/BA position is the same as a BA/PM position. The positions differ only in that the former is aligned with project management and the latter with business analysis. The skill profiles of the PM/BA sector and the BA/PM sector are very similar. This mirrored relationship does not hold for the PM/ba and BA/pm sectors. The PM/ba position is light on business analyst content while the BA/pm position is light on project management content.

This landscape was not revealed to me in a dream. It was the product of hours of trial-and-error graphic renditions. The first application was in the IT and business disciplines. In that application the sectors were IT, IT/b and IT/B. You get the picture. For nearly 30 years the IT industry has been severely criticized because its developers really didn't understand or appreciate the difficulty the business unit managers and business process owners were having in specifying what they wanted. How often have you heard an IT developer or systems analyst say: "I wish my clients could figure out what they want and clearly describe it. They no sooner give me a description than they change their minds. They have been known to make a decision and then reverse it later. When we finally deliver a solution to what they say they want, they aren't satisfied." Little did the developer know it, but in many cases it was not possible for clients to specify with any clarity or finality what they really needed. They could only guess at the requirements they needed and how those would be presented. At times the developers would suggest a feature or function that triggered an idea on the part of clients that later became an addition to the current solution. The problem ultimately resides in the processes used to gather requirements, analyze them, document them, write the project charter, plan and execute the project, and manage change. The processes did not live up to the realities of the project. The solution had eluded management for some time, and the processes that they would follow weren't exactly the ones needed to ferret out an acceptable solution.

> *I wish my clients could figure out what they want and describe it to me. They no sooner give me something than they change their mind. And when we finally deliver a solution to what they say they want, they aren't satisfied.*

The problem of wants versus needs is central to the contention between a business unit and its partners from the IT department. The BA can mitigate much of this problem.

Enter the agile movement in 1991 with the publication of "The Agile Manifesto" (Martin Fowler and Jim Highsmith, *Software Development* 9, no. 8 [August 2001]: 28–32). Agile projects did not have clearly defined solutions, and the developers knew that. That meant that complete requirements gathering was not possible during the scoping phase, and hence the traditional approaches for planning and scheduling a project were not a good fit. There are several agile approaches available now. They begin with an incomplete solution and through iteration discover the remaining parts of the solution.

"THE AGILE MANIFESTO"

We are uncovering better ways of developing [products] by doing it and helping others do it. Through this work we have come to value:

- *Individuals and interactions over processes and tools*
- *Working [products] over comprehensive documentation*
- *Customer collaboration over contract negotiations*
- *Responding to change over following a plan*

That is, while there is value in the items on the right, we value the items on the left more.

"The Agile Manifesto" ushered in a whole new approach for meeting incomplete requirements and accommodating change in collaboration with clients. In the intervening 20 years, as agile approaches to managing projects have been introduced into the PM and BA processes and practices, I have seen a marked change in focus. Meeting client needs and delivering expected business value are now the measures of a successful project. Meeting scope, cost, schedule, and resource constraints are nice to have, but they have nothing to do with business value.

From the client's perspective, the problem now looks like this. I've come to the conclusion, from 20 years' experience with client scoping sessions, that clients hopelessly confused their wants with their needs. What clients said they wanted was usually not what they needed. But they didn't know that and couldn't know that because no one had perfect knowledge of a solution. In clients' minds, they had a problem, and what they wanted was what they thought would solve their problem. They were putting forth the solution *as they saw it.* But their problem was undefined at the time they expressed their wants to the developers. This problem has a long history.

Starting in the 1950s, the computer was becoming a tool that could be used to automate processes like accounting, inventory control, salary, accounts payable, and accounts receivable. Clients chose not to explain their needs to developers, feeling that developers wouldn't understand because they didn't know anything about business processes. So clients chose to describe what they wanted, thinking that it would solve their problem. In some cases the solutions might be right on. In other cases they were far from acceptable. Clients might have come to the realization that what they said they wanted was not what they needed, and they didn't know that until they had a chance to review the solution that was put before them. Obviously the process that they followed under the leadership of the development team was flawed. But don't think that the blame is entirely theirs.

Since the 1950s, often what clients said they wanted was not what they needed.

We are now in the world of the agile and lean organization. New technologies and creative applications of existing technologies are critical to the business processes and products of organizations that expect to remain competitive. IT professionals, PMs, and BAs are inextricably linked in a collaborative partnership to position their organization for success. Each needs to have at least a fundamental knowledge of the other two disciplines. The BA skill set includes facilitation skills, process mapping tool use, logical data modeling processes, and user interface design. A junior BA needs to have applications skills, and the senior BA needs to be able to creatively apply and leverage most of these skills. Soon we will begin to see the integration of other skills, such as cost/benefit analysis, business case development, change management, and business architecture.

From developers' perspectives, the problem looks like this. In their desire to satisfy the client, the development team (under the leadership of the PM) makes a good-faith effort to understand what the client wants. Detailed requirements are gathered and documented. Signatures affixed and projects begun. If the blame for project failure can be assigned to anyone, it may be that the PM assumed that client wants accurately reflected actual client needs. The PM seldom did anything to check the validity of this assumption. That behavior has to change!

From the PM's perspective, the problem looks like this: The naive PM assumes that what clients want is what they really need. The PM proceeds to plan the project and is surprised that, when the project delivers what the client wanted, the client is not satisfied. The experienced PM assumes that what the client wants is *not* what the client needs until proven otherwise. This is always the safer ground. All fingers should point to the requirements elicitation process as the way to align wants and needs and assure the delivery of a solution that meets needs will do the job. A root cause analysis is something the PM or preferably the client's BA should conduct. Then the PM can proceed with confidence that the project is delivering what the client needs. But it is too early to declare victory. The world changes and so do client needs. So there must be a process built into the Project Management Life Cycle (PMLC) to maintain the alignment of needs to any change of needs. Enter the BA. The BA will be the link between the client and the developer and work through the PM to complete that linkage.

If the blame for project failure can be assigned to anyone, it may be that the PM assumed that client wants accurately reflected actual client needs. The PM seldom did anything to check the validity of this assumption.

To further complicate the situation, the project management and systems development life cycles (SDLCs) being used didn't support the challenge that both parties faced, even though they might not have been aware of the support problem There was a serious deficiency between the two life cycles. It wasn't until the 1990s that the agile movement formally began and a resolution of the life cycle misalignment emerged.

The BA is put immediately between the client and the PM and performs an essential role that perhaps only this person can do effectively: define, document, prioritize, and manage client requirements. If requirements are defined, documented, and maintained correctly, the misalignment of wants and needs should go away. If this process is effectively designed and executed, I believe we can make a big impact on project success and client satisfaction. The Conditions of Satisfaction (COS) process described in Chapter 1 is a best practices process that I designed to do just that. I have used it for over 20 years and wouldn't think of scoping a project without the COS.

So my original interest in developing this landscape was for application to the IT profession. Many of the thought leaders in systems development believed that systems developers needed to learn more about the business of their clients and that clients needed a better understanding of the systems development processes used by the development team. Lack of a common language was a major obstacle to effective communications and understanding between developers and their clients. In other words, both parties needed to know something about the other's discipline. The same is true of the PM and the BA.

So far we have defined the levels of the PM and BA landscape. Those are shown along the leftmost side of Figure 3.1 and were defined in Chapter 2. Understand that this structure is not fixed. It can be adapted to the needs of the organization. The number of levels can be adjusted to fit a particular position family. I expect that as the size of the organization increases, the depth and breadth of the PM and BA position family will grow accordingly.

The six sectors define the mix of the two disciplines from project management only (the PM sector) to business analysis only (the BA sector). Between those two extremes are sectors that mix both disciplines. Some PM positions require an elementary working knowledge of business analysis (the PM/ba sector) while other PM positions require an in-depth knowledge of business analysis (the PM/BA sector). On the BA side of the landscape are parallel positions where the BA needs an elementary working knowledge of project management (the Ba/pm) or an in-depth knowledge of project management BA/PM sectors). The two sectors at the middle are virtually identical from a skill and proficiency perspective. The only difference is one of organizational alignment. The PM/BA's primary focus is project management. The BA/PM's primary focus is business analysis. The six sectors are probably a fixed part of the PM/BA landscape. The number of position levels is the variable.

For example, Wal-Mart has about 285 independent client groups among its project management and business analyst professionals. They are all housed in the IT department. Wal-Mart has over 10,000 professionals in its IT department who are spread across these client groups. Each client group contains several business and technical professionals whose position scopes extend across the needs of their client group. Each client group is staffed to provide all of the IT, business architecture, business process, and project management support that client group needs. I don't think anyone knows how many different positions Wal-Mart has in its PM/BA position family. Whatever the number of different positions might be, it probably exceeds 1,000! These positions are spread across the six sectors. Wal-Mart has simplified the problem of naming by calling all of them "associates." But keep in mind that those positions encompass IT, project management, business analysis, and subject matter experts (SME) in the product/process of the client group. Each sector is densely populated with positions. The landscape is big, and it is comprehensive! When I was actively consulting with Wal-Mart's IT department, the IT staff often said that they typically knew more about the business processes of their clients than the clients knew. In the cases I encountered, I tended to agree with their opinion.

Figure 3.1 should be interpreted as robustly as possible. First of all, keep in mind that the cell structure can change to fit the position family needs of an organization. Then this landscape includes every PM and BA position that exists now or will ever exist for that organization. For the purposes of this book, I use only eight position levels, ranging from Team Member to Director. An organization of any size or complexity will probably define additional position levels. An individual's career path in PM, BA, or some combination of the two begins in one of the six vertical sectors at the Team Member Level. A vertical sector defines a position that contains some mix of PM and BA content. So an entry-level position is at the Team Member Level in one of the six sectors. As individuals gain in experience and acquire new or improved skills, they move up through higher-level positions. Those positions will consist of PM or BA responsibilities or some combination of the two.

Each vertical sector represents some mix of PM and BA responsibilities. The extreme left and extreme right sectors identify those positions that have either pure PM or pure BA content. These positions require skills in only one discipline, and their assignments have no such requirements or expectations in the other discipline. The people who occupy positions in these sectors may have skills and competencies in the other discipline (that probably is true in the majority of cases) but there is no requirement in either of these sectors that that is true. The two disciplines do not have independent skill sets; hence there will always be some skill overlap. The focus here is on the positions and the skill requirements of those positions, not the people and their skills. All of the sectors in between contain positions having some combination of PM or BA content. Over the chapters in this book we will look inside each of the 48 cells in Figure 3.1 and define the PM and BA positions and skill profiles in detail.

So far I have only talked about positions, and that is how the landscape should be interpreted. Each cell is defined by a level (Team Member through Director) and some mix of PM and BA content, which defines the sector that the cell occupies. Each cell can contain one or more positions. All of the positions in an organization that contain PM or BA or some combination of the two can be found in one of the cells in this landscape. Obviously all of the positions that occupy a specific cell have a lot of skill/proficiency commonality. There is a minimum skill/proficiency profile for each cell, and so to occupy a position in that cell, an individual must meet or exceed the minimum skill profile for that cell. The position itself may require additional skills and/or proficiencies. So in a multidimensional space defined by skills, you can think of the positions in a cell as being ordered from those positions that just meet the minimum skill/proficiency profile (positions in the lower left corner of the cell) to those that almost meet the minimum profile of an adjacent higher-level cell (positions near any of the other three corners of the cell). This ordering will be important for career path planning systems. Chapter 4 describes the skill/proficiency process used and Appendix A contains the minimum skill/proficiency profile for every one of the 48 cells in the landscape.

I've constructed Table 3.1 to include some examples of specific positions within the cells. The table is not exhaustive of the possibilities but is illustrative only. The position titles are not necessarily tied to any company but reflect my intentions for developing career paths for both PMs and BAs and hybrids of the two disciplines.

In Table 3.1, APM is an abbreviation for Agile Project Management and TPM is an abbreviation for Traditional Project Management.

I've taken some liberties in naming positions in the spirit of beginning a discussion about career paths, especially in the BA sectors of the landscape. There are discussions in the literature regarding Business Analyst and Business Architect position titles. These seem to define one version of a BA career path. I see career opportunities for BAs as managers of a specific business process—Order Entry and Order Fulfillment Process Manager, for example. To fit at the Senior Manager Level, the business process would have to be a more significant part of the business—Supply Chain Process Manager, for example. My guess is that when an organization populates the PM/BA landscape with all of its positions, the organization will see a trend

TABLE 3.1 Examples of Typical or Hypothetical Position Titles

Position Level	PM	PM/ba	PM/BA	BA/PM	BA/pm	BA
Director		PMO Director	Project Portfolio Director	Chief Process Officer	Director BA Management	
Senior Consultant		Performance Metrics Consultant	Senior PMLC Methods Consultant	Senior Business Architect Consultant	Senior Business Analysis Consultant	
Consultant	Estimation Consultant	PM Career and Professional Development Consultant	PM/BA Curriculum Consultant	BA/PM Curriculum Consultant	BA Career and Professional Development Consultant	Supply Chain Process Consultant
Senior Manager		Senior Project Manager	Program Manager	Senior Business Architect	Supply Chain Process Manager	
Manager		Project Manager	APM Project Manager	Client Relations Business Architect	Order Entry and Order Fulfillment Process Manager	
Associate Manager		TPM Project Manager			Business Analyst	
Task Leader	Acceptance Test Supervisor					Business Process Analyst
Team Member	Acceptance Test Technician					Business Analysis Technician

toward an increasing number of position titles in the PM/BA and BA/PM cells as you move up the career paths. This should plant a career development idea in the minds of both the PM and the BA: Moving to positions toward the middle sectors (PM/BA and BA/PM) is a wise strategy.

PM and BA Sectors

Positions that have only pure PM or pure BA content do not occur that frequently. In fact if you look at the Project Management Body of Knowledge (PMBOK) and Business Analysis Body of Knowledge (BABOK), you will see that there are tasks performed by both professions included in both documents. Every PM performs some BA tasks, and every BA performs some PM tasks. The two professions are inseparable from a skills standpoint. Positions at the Professional Level and Executive Level will always include tasks from both disciplines. So the PM and BA sectors at the Associate Manager, Manager, Senior Manager, Consultant and Senior Consultant Levels are not well populated with positions.

The PM and BA sectors exist mostly for the sake of defining a complete landscape, even though only a few positions are found in these sectors. These positions are most likely to be found at the Staff Level, as Table 3.1 illustrates. People occupying Staff Level positions as Team Members typically have 0 to 18 months' job experience in PM or BA. At entry they may have no skills in either PM or BA. Early in their employment they will receive introductory training in one or more PM or BA tools and use those tools to fulfill their job assignments. As they gain in experience and move to a Task Leader position, they will acquire a broader set of skills that may include elementary tasks in one or both disciplines. For example, BA Task Leaders will need to have some skill at scheduling the work of those they may be supervising. Scheduling the work assignments of the Team Members under their supervision is an elementary project management skill.

PM/ba and BA/pm Sectors

These two sectors are densely populated with positions that have content in both disciplines. If you examine the PMBOK and BABOK standards, you will see that the traditional PM has roles and responsibilities that are the primarily the province of the BA. These are mostly in the Requirements Management and Change Management areas. Similarly, the BA has roles and responsibilities that are primarily the province of the PM. These are mostly in the Scope Management and Implementation areas. The professionals who occupy these positions routinely practice both disciplines. That does not imply that these positions include dual responsibility for PM and BA on the same project. Rather the positions require individuals to use their skills in the support discipline to assist with execution of tasks in their primary discipline. For example, a BA involved in a process improvement project will have to create iterative plans that test ideas for process improvement. This will require prioritizing suggested ideas for process improvement, allocating those ideas to iterations, creating a work breakdown structure (WBS) for each iteration, and scheduling the required resources. These are simple project management skills that a BA/pm at the Associate Manager Level will need to master.

PM/BA and BA/PM Sectors

In the middle sectors are the PM/BA and BA/PM positions. These are the more contemporary positions that I have been referring to in the previous chapters. I'll take some risk here and suggest that the PM/BA and BA/PM sectors offer the most career opportunities, the PM/ba and BA/pm sectors lesser career opportunities, and the PM and BA sectors the least. As professionals move up through position levels, they will find more opportunities if their skills profile contains both disciplines. So the two middle sectors are where the best career growth opportunities exist. The Professional Level individuals who occupy positions in these middle-sector cells are fully qualified to manage projects and manage business analysis engagements either separately or merged into the same project. The skills content and proficiency levels of the positions in the two sectors are equivalent. That is, a PM/BA Senior Manager position has the same content and skill/proficiency profile as the BA/PM Senior Manager. The only difference between these two Senior Manager positions is their primary orientation. I believe that there are needs for PM/BA and BA/PM professionals and major untapped career opportunities available for them. To date the two positions types have been kept separate. For the first time, in this book, I want you to seriously consider these dual-disciplined professionals and their place in the business world. I believe that the seat at the strategy table will be accessible to professionals who progress in the two disciplines through the PM/BA or BA/PM career paths. A lot has to happen in the development and definition of both professions before that becomes a reality. I ask you to keep an open mind, as many of these developments are a work in process. The rows of this landscape refer to the eight levels in the position family. These are the same position levels shown in the generic landscape of Figure 2.1.

A seat at the strategy table will be accessible to professionals who progress in the two disciplines through the PM/BA or BA/PM career paths.

At the Staff Level there are two positions: Team Member and Task Leader. Team Member is the Entry Level position. The skill/proficiency requirements in PM and BA for the positions at the Staff Level are general skills with minimal proficiency requirements. These positions include initial training in a few tools, templates, and processes that prepare the individual for simple and closely supervised assignments in either PM or BA. As they gain experience they will move up to the Task Leader Level, where they will be qualified to supervise the work of one or more Team Members assigned to the task. A Task Leader is generally a working supervisor position. This is the Entry Level position to people management.

At the Professional Level there are five positions. The lower three are manager positions. The upper two are consultant positions. Associate Managers are qualified to manage small, simple projects or major deliverable parts of larger projects. (Remember we are defining projects from the broadest perspective.) Through experience they progress to the Manager Level. They are now qualified to manage even the most complex projects. The Senior Manager position manages large projects and programs and often has Managers and Associate Managers reporting to them. For those professionals who prefer to continue their career growth, they can follow an individual contributor path on the Consultant Track, which consists of two positions, or move directly to higher-level management positions at the Executive Level. The Consultant is an SME for some business processes, including project management. The Senior Consultant is a position level attained after experiences that broadened and deepened the consultant's process knowledge.

At the Executive Level there are two director position types: Process Director and Practice Director. The Process Director is a senior executive responsible for all activities associated with a single business process. These will generally be major business processes. The Practice Director is a senior executive responsible for all activities associated with managing the people side of projects. The two Director Level positions lead to a variety of C-Level positions at the enterprise level and are outside the scope of the PM/BA landscape.

This landscape depicted in Figure 3.1 is the playing field for the career and professional development of the BA and the PM. There are 48 distinct cells in this landscape, and all BA, PM, and dual-discipline position configurations fall somewhere in it. It therefore can be used for career planning and professional development. Each cell will contain one or more position titles, and each position title will have a skill and competency profile defined for it. An individual's career history can be represented by some sequence of position types within a sequence of connected cells through this landscape. An individual's professional development plan is represented by some planned sequence of connected cells through this landscape. Chapter 7 offers detail on using Figure 3.1 for career and professional development planning and execution.

Organizational Placement of the PM and BA

Once upon a time, systems analysts were housed in the IT department and divorced from clients and hence were viewed as ineffective in understanding and meeting client needs. If they were housed in the client organization, they could quickly lose touch with their technology and were seen as ineffective by their IT comrades. The same thinking has invaded the PM and BA world. I am not aware that there has ever been a rigorous analysis of this topic. Although that is outside the scope of this book, I think it is incumbent on every contemporary organization to perform due diligence on what contributions it expects from these two disciplines and use that perspective to decide on organizational placement of the PM and BA. That will inevitably lead to an integration of the two disciplines as depicted in Figure 3.1. Most organizations already have a CIO Executive Level organization in place, and the project management organization is usually highly placed at the Director Level within the office of the CIO. That structure provides for the integration of the PM and IT professionals. So how does a BA function in this structure? One approach is to create a business unit that parallels the PMO but is focused on business analysis. Call it a Business Analysis Office (BAO) and put it at the same organizational level as the PMO. Ideally these two offices will report directly into the Office of the CIO. Another alternative is to expand the current PMO to include business analysis. I've named this the Business Process, Project, Program and Portfolio Management Office (BP^4SO). See Chapter 7 for the detailed discussion. Yet a third alternative might be to create two types of BAs: BA generalists and BA specialists. The BA generalists are housed in a BAO and provide consulting and training support to any business unit. The BA specialists are housed in the client organizations and as SMEs provide direct business analysis support to the processes of that client organization.

In the final analysis, each organization is unique, and its solution to the PM and BA integration into its structure is unique as well.

Putting It All Together

Obviously this is a work in progress. It has been done for the IT profession but not for the BA/PM professional (or PM/BA if you prefer). This book provides the initial use of the landscape for the PM and BA professions. Much remains to be done. I would certainly like to hear your thoughts on the BA/PM professional or PM/BA professional, if you prefer. The next challenge recognizes that PM, BA, and IT are converging and that a successful professional will need some combination of skills in each of the three disciplines. And so the two-dimensional project landscape of Figure 3.1 will have to be replaced by a three-dimensional project landscape. That is also a work in progress. Email me at rkw@eiicorp.com. I'm sure we could have a lively exchange. I promise to respond personally to every email.

CHAPTER 4

Project Manager and Business Analyst Skill Profiles

This brief chapter defines the skill/proficiency profiles of the Project Manager and Business Analyst positions shown in the landscape of Figure 3.1. The actual proficiency-level data is quite voluminous. So as to not disrupt the continuity and flow of ideas in this chapter, the skill/proficiency data has been relegated to Appendix A. The data reflects my own opinion but to the best of my ability is consistent with the published literature. I put it forward as a first attempt at defining a complex skills profile for BAs and PMs. My approach is to define the minimum skill/proficiency profile for cell membership of people and/or positions.

There are 48 cells in the position landscape described by Figure 3.1. Every cell is further defined by the minimum skill proficiency profile required for every position that is classified in the cell. Some cells may actually be empty. For example, a position at the Director Level in the PM sector is not too likely. They would have to have at least a minimum of business analysis proficiency. Nevertheless, I will provide a skill/proficiency profile for the sake of completeness. In order to occupy a position in a cell, the individual must meet the minimum skill/proficiency profile for that cell. There may be circumstances where people are assigned a position in a cell whose profile they do not meet. Specific positions within a cell may require more than the minimum proficiencies for that cell or additional requirements outside of the specific skills required for that cell. This chapter defines an established model I have adapted to establish those skill proficiency profiles. These are minimum profiles. An organization may choose to raise those minimum proficiencies for all positions in a cell or for specific positions in a cell. The actual skill/proficiency data, given in Appendix A, and should be interpreted as a template to be used as a starting point for any organization.

To the extent possible, I want to respect the extensive work done by both the Project Management Institute (PMI) and the International Institute of Business Analysis (IIBA). Neither the Project Management Body of Knowledge (PMBOK) nor the Business Analysis Body of Knowledge (BABOK) contain skills profiles. Those documents are standards guidelines and nothing more. Skill profiling is out of scope for both documents. To the best of my ability, I tried to be consistent with those two documents in the decisions I made. I bear the final responsibility for any errors or omissions.

BA and PM Proficiency Model

My goal was to find or define a metric that can be used to define the skills profile for every one of the 48 cells in the PM/BA landscape shown in Figure 3.1. Having or not having a particular skill is too simplistic. We need a scale that meets these five criteria:

1. **Simple enough to quickly assess skill/proficiency levels**. PMs, BAs and perhaps their managers are the target audience for this proficiency model. The PMs and BAs can be expected to use it twice a year at most and so they never get enough use to retain any learning. So it has to be simple. The managers of these professionals will use it once or twice a year for every one of their PM or BA subordinates and so it needs to be quick. Fatigue and inattentiveness can set in quickly if there are too many assessments to be done over a short period of time.
2. **Robust and applies equally to every skill in the PM and BA profiles**. The architecture and contents of the assessment instrument must be adaptable to organizational needs. Skills will need to be added. Proficiencies will need to be changed. The proficiency structure and values need to make sense as different positions are compared.
3. **Distinguishes among knowledge, experiences, and mastery**. Knowledge precedes experience and experience leads to mastery. The various proficiency levels must have some easy way of distinguishing among each level. The question reduces to objectively knowing the differences based on behavioral patterns of the individual. Observable behavior patterns will be important to the PM or BA manager because they do not have intimate knowledge of the individual's actual skill.

4. **Distinguishes levels of expertise ranging from novice through expert**. The model must do more than give a yes/no answer to whether an individual possesses a skill. Having knowledge of a skill does not imply the ability to use the skill, and being able to use the skill does not imply mastery of the skill.
5. **Objective, repeatable, and intuitive**. The determination that an individual possesses a specific level of proficiency must be as objective as possible. That allows comparisons over time of that individual's progress toward experience or mastery. It also allows comparisons between individuals, which will be useful for PM and/or BA assignments.

Selecting an Assessment Approach

There are several ways to proceed. I could define a metric just for this book that meets these five criteria. I'd rather not reinvent the wheel since there are several models already in use and a new model would not have the credibility and research background of an established one. I reviewed the Dreyfus Model (Stuart E. Dreyfus and Hubert L. Dreyfus, "A Five-Stage Model of the Mental Activities Involved in Directed Skill Acquisition," California University Berkeley Operations Research Center [February 1980]). It is an established model used by the IIBA in its recently published BA Competency Model (IIBA, *A Guide to the IIBA Business Analysis Competency Model, Version 1.0* [Marietta, GA: Author, 2010]). The Dreyfus Model has six stages of skill proficiency:

1. Novice
2. Advanced Beginner
3. Competent
4. Proficient
5. Expert
6. Proficient Expert

This is a perfectly good model and is in wide use. So far the IIBA Competency Model has been applied only at the Proficient Level. Its use of the Dreyfus Model is perfectly valid, but it does not meet my five criteria. For example, to be at the Proficient Level, the individual must meet several criteria specific to the skill being assessed. Every skill has a different set of criteria. That places a huge burden on the assessor (individual or manager). The document describing the criteria becomes a crutch for assessors, and they cannot do an assessment without having that document at their side. The document is 41 pages long, and that is just for one of the six proficiency levels of the Dreyfus Model. To meet my criteria for intuitiveness, the metric must be robust. That means there should be criteria that are the same at a given level of proficiency regardless of the skill being assessed. This is achievable. I have done it for project management and I can extend the same concepts to business analysis. That minimizes the need for constantly having to reference lengthy documents to do an assessment. With some practice, the assessment of proficiency becomes intuitive.

The only model that I have found that meets all five of my criteria is Bloom's Taxonomy (Benjamin S. Bloom, *Taxonomy of Educational Objectives, Handbook 1: Cognitive Domain* [Boston, MA: Addison-Wesley, 1956]). Bloom's Taxonomy is a six-level proficiency model, just as the Dreyfus Model is. The two are quite similar in purpose but different in application. Bloom's Taxonomy is very well known and has been used extensively in training and learning models. I originally adapted it in 1990 to fit my needs for project management skill/proficiency assessment and have used it successfully with my clients since then. It does a particularly good job at assessing proficiency at the knowledge, experience, and ability levels. With the publication of this book I am expanding its application into the business analysis discipline and the PM/BA landscape defined in Figure 3.1. My adaptation consists of the 25 behavioral characteristics from Bloom's Taxonomy spread across a 6-level proficiency scale, as shown in Table 4.1. The proficiency level of an individual with respect to a specific skill is determined by comparing observable behavior to a set of 25 defined behaviors. That same set of 25 behaviors is used for every skill. So assessing an individual's proficiency is a repeatable process for every skill. The learning curve is very short.

Bloom's Taxonomy has six levels of proficiency:

Knowledge (Levels 1 and 2): In Bloom's Taxonomy, these are defined as Knowledge and Comprehension.

Experience (Levels 3 and 4): In Bloom's Taxonomy, these are defined as Application and Analysis.

Ability (Levels 5 and 6): In Bloom's Taxonomy, these are defined as Synthesis and Evaluation.

The proficiency level of an individual with respect to a specific skill is determined by comparing observable behavior to a set of 25 defined behaviors. That same set of 25 behaviors is used for every skill. So assessing an individual's proficiency is a repeatable process for every skill. The learning curve is very short.

TABLE 4.1 An Adaptation of Bloom's Taxonomy (Cognitive Domain) to PM and BA Skill/Proficiency Assessment

Proficiency Level	Descriptive Label	Observable Behavior K = Knowledge E = Experience A = Ability
Level 0	I don't know anything about it.	
Level 1	I can define it.	K1.1 Familiar with the terminology K1.2 Understands the basic concepts
Level 2	I understand what it can do.	K2.1 Knows how it is used K2.2 Can explain key issues and benefits K2.3 Understands organizational relevance
Level 3	I have limited hands-on experience.	E3.1 Has a working knowledge of basic features and functions E3.2 Aware of relevant standards, policies, and practices E3.3 Requires assistance and supervision E3.4 Can apply it in a limited (homogeneous) environment
Level 4	I have extensive hands-on experience.	E4.1 Knowledge of operational issues and considerations E4.2 Understanding of benefits and drawbacks E4.3 Working knowledge of relationships and integration E4.4 In-depth knowledge of major features, functions and facilities E4.5 Awareness of usage in other environments E4.6 Can work without assistance or supervision

TABLE 4.1 *(Continued)*

Proficiency Level	Descriptive Label	Observable Behavior K = Knowledge E = Experience A = Ability
Level 5	I can adapt it to a variety of situations.	A5.1 Theoretical background and understanding A5.2 Expertise in all major features, functions, and facilities A5.3 Experience in multiple environments (heterogeneous) A5.4 Knowledge of and contribution to "best practices" A5.5 Ability to consult and coach others
Level 6	I am recognized as an expert by my peers.	A6.1 Extensive experience in multiple/complex environments A6.2 Industry and marketplace perspective A6.3 Historical and future perspective A6.4 Influencing wide or high-impact decisions and initiatives A6.5 Leadership on architecture, policies, strategy, and "best practices"

In order to be proficient at, say, Level 4 for a particular skill, there must be visible evidence that the behavioral characteristics E4.1 through E4.5 have been observed in the person's work habits. Continuing the same example, suppose the individual also has given evidence of having used behaviors A5.1 and A5.2 for the same skill. In other words, the individual has demonstrated proficiency in 2 of the 5 behaviors needed to establish Level 5 proficiency. That doesn't mean that the person has attained proficiency at Level 4.4 for that skill. Until the individual has demonstrated proficiency in all 5 behaviors that describe Level 5 proficiency, he or she is still proficient at Level 4. Knowing which behaviors the individual has not yet demonstrated will inform him or her as to training or experience still needed in order to attain Level 5 proficiency. Bloom's Taxonomy is cumulative. To be proficient at Level 4 implies that the individual is also proficient at Levels 1, 2, and 3 for that skill.

The simplicity in this adaptation is that the same 25 behavioral traits are used for assessing every skill. I have found that with a little practice, the assessor gets more adept at using the model and it gets easier to make proficiency level decisions. The process really does become intuitive after the assessor has a little experience using the model. I've had the occasion to do some face-to-face assessments. I know the skills list for the PM so well that I can just have a conversation for a few minutes and begin to get a very accurate reading on an individual's proficiency levels.

I know that using this robust approach may not be detailed enough for some applications. If that is the case, then I would recommend adopting the IIBA approach and using the Dreyfus Model. You have to decide how much effort you are willing to commit to creating the detailed content.

There are two major benefits from having collected this skill/proficiency data:

1. Ability to classify a position into a specific cell
2. Ability to classify an individual into a specific cell

Having positions classified into the 48 cell structure and individuals classified into the same cell structure makes it possible to design and implement several other management programs:

- Skill assessment process
- Career and professional development program
- Project portfolio management process
- Training curriculum design, development, and scheduling
- Resource planning process
- Resource management process

Let's take a look at each application.

Skill Assessment Process

A simple process for skill assessment by the individual or manager can be developed. I have a web-based tool that has been used for years by my clients' PMs to self-assess their skills/proficiencies as input to their professional development planning process and to the training manager to plan and schedule training courses. The assessment is usually done annually so that progress against a development plan can be measured by calculating the proficiency changes from the previous assessment. My clients' experiences suggest that this approach and level of detail is appropriate for most needs. The development of more detailed data that is specific to tool, template, or process is not needed. It is implied based on the specific products in use in the organization. For example, if a specific project management software system is in use, the assessment of proficiency level can be made using generic behavioral statements, knowing that it is done with reference to the specific software package being used.

If the organization changes software packages, it will not affect the skill assessment process. It is robust with respect to specific products. So, for example, if your organization uses MS Project and changes to some other vendor's project management software, the assessment does not have to be changed.

There are three different variations of the use of a skill/proficiency assessment. They are listed next.

Self-Assessment

In my experience, self-assessment has been the most commonly used variation. Here individuals assess their own proficiencies and present the results to their career advisor. If this is the second or some subsequent assessment, then there are two consecutive assessments to compare. The assessment is used to build the next Professional Development Program (PDP).

There is concern that individuals will bias their assessments, either claiming a higher or lower proficiency than is actually the case. Their reason for claiming a lower proficiency is so that they can plan on registering for off-site training and get to some exotic location for fun and games rather than actual training. In my experience this is a myth with no basis in reality. Their reason for claiming a higher proficiency is to get promotions and salary increases. Again I haven't seen any evidence that this is happening.

If bias is a concern, then I suggest that individuals and managers come to an understanding of the purpose of skill assessment and the importance of making it as accurate as possible.

Manager Assessment

Manager assessment is a variation but has not been a popular choice among my clients. Managers usually are not in a position to directly observe behaviors and make a fair assessment of proficiency levels. In those cases where they can, they discuss their assessments with the individual and make any adjustments that the individual has been able to satisfactorily defend.

Self-Assessment and Manager Assessment

The self-assessment and manager assessment has been the most valuable variation, even though it is the most labor intensive of the three. Once both assessments are complete, individuals and their managers meet and compare results. Those skills for which there is a disagreement between the two proficiency assessments are discussed, and the variance is removed by making the appropriate change to the proficiency level. The individual's assessment becomes the agreed assessment with the variances resolved in an updated version.

Career and Professional Development Program

The skill/proficiency profiles of the current PM and BA can be matched against the PM BA landscape skill/proficiency of each cell to determine the cell the individual best fits. That will be the highest level cell for which the person's profile exceeds that of the profile that defines the minimum proficiency levels of every skill that defines that cell.

There is a natural ordering of cells within the PM/BA landscape. This is important to defining a career path. For any two adjacent cells on the PM side of the landscape, the one that is at the same position level and is farthest to the right or at a higher position level in the same sector is the higher cell. For example, the Senior Manager Level PM/BA cell is higher than the Manager Level PM/BA cell, and the Manager Level PM/BA cell is higher than the Manager Level PM/ba cell. For any two adjacent cells on the BA side of the landscape, the one that is at the same position level and is farthest to the left or at a higher position level in the same sector is the higher cell. For example, the Senior Manager Level BA/PM cell is higher than the Manager Level BA/PM cell, and the Manager Level BA/PM cell is higher than the Manager Level BA/pm cell. For cells in different sectors at different position levels, there is no definable ordering. So cell ordering implies position ordering. If cell A is higher than cell B, then every position in cell A is higher than every position in cell B, and the proficiency

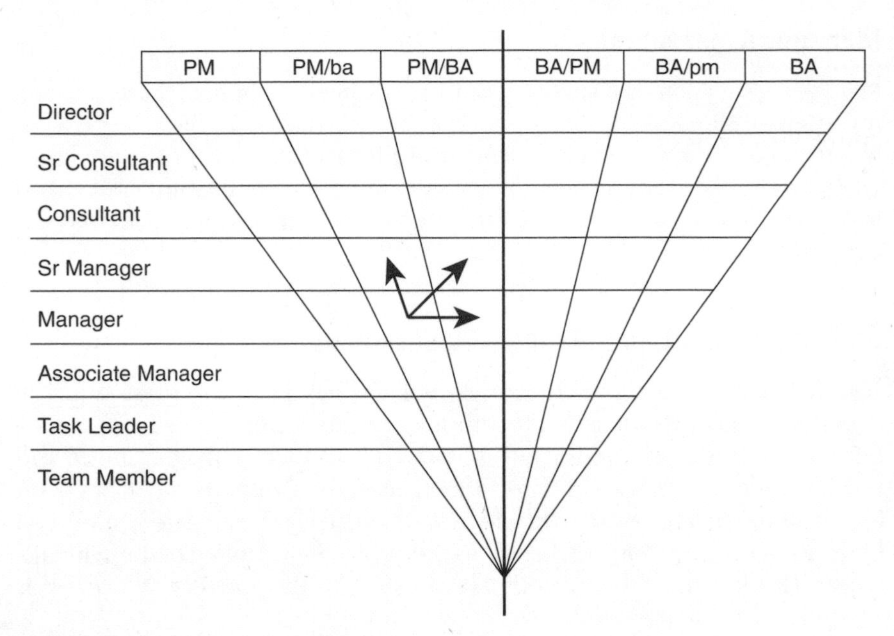

FIGURE 4.1 Natural Ordering of Cells

level of every skill in cell A is higher than or equal to the proficiency level of every skill in cell B. To add a little more clarity to this ordering scheme, I've put an example of ordering into the landscape and displayed it in Figure 4.1. The direction of the arrows indicates the higher-level cells.

This natural ordering of cells is also useful for career advancement. For this example, the individual currently occupies a position in the Manager PM/ba cell. This person has three choices for moving "up" in his or her career: to Senior Manager PM/ba, to Manager PM/BA, or to Senior Manager PM/BA. All three of these choices denote valid career path progress to a higher-level position. The one the person chooses to follow is a function of the individual's career goals, available training, and opportunities for advancement.

The actual movement along a career path is more a function of the opportunities that are available and when they are available than it is about an individual's personal desire for advancement. Wanting to advance is the prerequisite, but the opportunities that are available are the controlling factor. The manager has to assume that the individual's career goals may be in another part of the company or even another company. In either case, the manager should be supportive. In particular, the individual will take advantage of those opportunities when and if they occur. The three examples shown next give a taste of how the landscape will be used for career planning and advancement. A more detailed discussion and several example career paths can be found in Chapter 7.

Some Short-Term Career Development Paths

Here are a few examples of short-term career paths. In Chapter 7 we'll discuss longer-term career paths and some of the implications to PM and BA professionals.

FROM MANAGER PM/ba TO SENIOR MANAGER PM/ba This career path progress is mostly the result of seniority. That is, the PMs spent time in grade and gained experiences that qualify them for the next level of project management. Along with that seniority they will have managed larger or more complex projects and acquired the skills to do so along the way. They will qualify for a position where they can expect to manage other PMs in large projects or programs.

FROM MANAGER PM/ba TO MANAGER PM/BA Many PMs will have recognized that the more they know about business analysis, the more opportunities they will have to manage projects that draw on business analysis skills and proficiencies. They will qualify for such positions more as a result of business analysis training. Some of that training will be on the job, where they have been involved in projects that challenge them to improve their business analysis proficiencies. Perhaps they will have an opportunity to work alongside a senior-level BA. Their involvement in such projects may be more supportive because their goal is to learn new skills in business analysis rather than in project management. Some of those skills may deal with increasing their proficiency in skill areas that focus on the collaborative relationships between PMs and BAs. In general, PMs and BAs are going to improve their collaborative skills by working together.

FROM MANAGER PM/ba TO SENIOR MANAGER PM/BA Qualifying for career moves like these is more encompassing and challenging than the previous two cases. The advancement requires experience acquisition and skills acquisition at the same time. It is more aggressive than a two-step process like:

1. Manager PM/ba to Manager PM/BA to Senior Manager PM/BA
2. Manager PM/ba to Senior Manager PM/ba to Senior Manager PM/BA

The two-step process generates the same results as the more aggressive one-step career movement.

Calculating a Skill Gap

A skill gap is a metric that measures the skill separation between your current skill profile and the skill profile of another position. That other position is usually the next step in career progression in your development plan.

Let's take a very simple example to illustrate the skill gap formula that I use. Suppose that the position skill profile consists of only four project management skills (PM1 to PM4) and three business analysis skills (BA1 to BA3). Table 4.2 lists the skill profile for your current position and some target position.

Note that there is a skill gap only for those skills whose current proficiency level is below that required by the target position. In other words, if the individual's current proficiency level exceeds that of the target position for a given skill, then the person's proficiency skill gap for that skill is 0. The skill gap for PM2, PM3, and BA3 are 0. The individual has to improve proficiencies in PM1, PM4, BA1, and BA2 in order to be qualified for the target position. The formula for geometric distance between two objects says to sum up the squared differences on each dimension and take the square root of that sum. Squaring the differences eliminates the effect of negative differences. That is the geometric distance between the two objects. For our skill gap metric, I am altering that definition by excluding the negative differences (that would be where the current proficiency was greater than the target proficiency). That allows me to simplify our Position Skill Gap Formula to be the sum of the individual proficiency skill gaps. For this example, the Position Skill Gap is 7 (2+2+2+1). There is no need to take the square root. The target position is 7 units away from the current position. While this is not the geometric distance between two skill profiles, it is a pure number that can be compared to other skill gaps. Using this metric, you can order all feasible target positions from nearest to the current position to farthest from the current position. Those that are close

TABLE 4.2 Calculating the Position Skill Gap

Skill	Current Proficiency	Target Proficiency	Proficiency Skill Gap
PM1	3	5	2
PM2	4	4	0
PM3	4	2	0
PM4	2	4	2
BA1	3	5	2
BA2	4	5	1
BA3	2	1	0

require less effort to reach while those that are far away require a more aggressive skill enhancement effort.

People's skill profiles place them in a cell in the PM/BA landscape. If they know their short-term career path, they know the next cell they would like to qualify for entry. That cell will have a minimum skill profile required for entry. A skill gap can be calculated and a plan developed for resolving the skill gap.

NOTE: Other than defining the highest-level position a person can occupy, the cell in which an individual is classified has little to do with the cell to which he or she is assigned.

An individual's cell does identify the recommended cell in which the PM or BA assignment should be found, but it is not a hard-and-fast rule that the person will be assigned to a position in that cell. Actually, the cell defines an upper limit on the assignments a person is qualified to fill. There are a number of variables that will affect the final assignment. Just because I am a Senior Manager PM/ba, it doesn't mean I will be assigned a position in that cell. First, there must be a project which needs such a PM.

Project Portfolio Management Process

Projects that are already proposed and in the pipeline for inclusion in the project portfolio will each have a resource requirements list for PM and BA positions by cell. This is the initial proposed list, but it may be different from the staffing they actually obtain. If these needs are defined in terms of the PM/BA landscape, the portfolio manager will have a sense of the demand for PMs and BAs by level and by sector. This is the demand side of the equation for project approval and staffing.

In most cases the demand will exceed the supply, and some prioritization rules for proposed projects will need to be employed to bring demand within the constraints of supply.

The supply side of the equation for PM and BA staffing is the landscape populated with PMs and BAs whose skill/proficiency profiles have placed them in a cell and therefore identified the projects they are qualified to staff. So we have the PM/BA landscape populated with professionals who are qualified to manage projects, and we have a prioritized list of projects to be staffed. The Graham-Englund Model shown in Figure 2.3 in Chapter 2 can be employed to meet the staffing requirements for PMs and BAs. A simple deployment of the model would be to work down the prioritized list making staffing assignments until you encounter a project for which the PM and/or BA requirement cannot be met from the available resources.

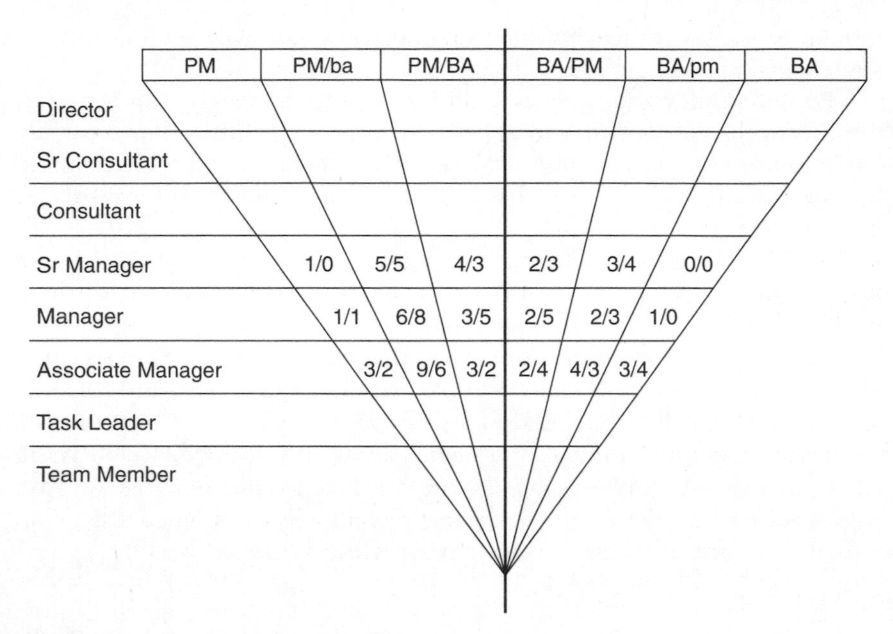

FIGURE 4.2 Supply versus Demand for PMs and BAs

This is a very simple rule to follow, but in practice several changes will be considered in order to staff more projects. These include changing:

- Project proposed scope
- Project proposed schedule
- Project approach to incremental delivery
- PM and/or BA staffing requirements

Thus, the skill/proficiency matrices can be used to populate the PM/BA landscape shown in Figure 3.1. On one hand, the landscape will display the supply of PMs and BAs in the organization. On the other hand, the landscape will display the demand for PMs and BAs based on the proposed projects. For discussion purposes, a simple example is shown in Figure 4.2.

The notation used in the figure is supply/demand. For example, the ratio 6/8 in the PM/ba Manager cell says that there are 6 PM/ba Managers on staff and there is a demand for 8 such professionals for the proposed portfolio. We don't know if any of them are available, but for the sake of the example, let's assume these numbers reflect availability. Availability data can be incorporated easily.

What might be done to cover this gap? There are always options, and there will be data available to help resolve the gap. For this example. some of the options are:

- Adjust the proposed projects to reduce or remove the gap.
- Change project schedules (i.e., delay one or more projects, change one or more project approaches to incremental or iterative).
- Rescope one or more projects.
- Hire contract PM/ba professionals.
- Assign staff from adjacent cells (i.e., from PM/BA Senior Manager, from PM/ba Associate Manager).

Project Proposed Scope

By reducing project scope, you may be able to relax a requirement for a PM and/or BA staff to utilize an individual in a cell that has a surplus based on the proposed project staffing requirements. For example, rather than requiring a Senior Manager PM/BA, the scope change may allow you to utilize a Manager PM/BA or Manager PM/ba. The original scope may be moved to a project proposal for the next portfolio cycle.

Project Proposed Schedule

Changing a proposed schedule is a practice that is already used quite frequently, although it should be done with caution. Stretching the schedule of one project may allow you to accommodate another project's schedule. It is easily done, but it has two effects you need to consider. First, it now gives that PM or BA resource another project to build into their schedule. You have now created a dependency between two projects where that dependency did not exist prior to the schedule change. A delay in one project may adversely affect the other project's schedule. So the risk to your successor project goes up. Where will you place your priority? Second, if the schedule changes affect a PM's workload and the person is now managing two projects, you may have burdened him or her beyond what is reasonable. Now the risk increases for each of the projects the PM is assigned to manage.

> *Most experts suggest that an individual's workload should not include more than four projects and only one of those four should be at the management level.*

Project Approach to Incremental Delivery

Choosing an incremental release schedule for project deliverables has the advantage of getting to product to market sooner and process in place earlier without too much of an adverse effect on the original schedule. But there is another advantage. By coordinating the deliverables schedule between two projects, limited resources can be scheduled across more than one project at a time. For example, suppose both projects are process improvement projects. There is only one BA available for documenting the "As Is" and "To Be" business process. Pick the project whose "To Be" process has not been defined, and assign the BA to document its "As Is" process first. The client can begin considering what the "To Be" process should do. While that is taking place, the BA will document the "As Is" process for the other project. What you have done is create some downtime for the BA on the first process while she works on the second process. Just as in the previous section, however, risk will increase on both projects.

PM and/or BA Staffing Requirements

How often have you heard "We don't have a BA/pm Senior Manager. Can't you get by with a BA/pm Manager?" One of my past clients who happened to have the most sophisticated Human Resource Management System (HRMS) I have ever encountered asked those kinds of questions quite frequently. In order to fully utilize existing and available staff, the client asked PMs to take a look at their skill/proficiency requests and see if they couldn't relax them to create some staffing alternatives and development opportunities for lesser-skilled or experienced individuals. There will always be opportunities to do this. As a side benefit, you will also be creating an opportunity for a team member to learn new skills or improve existing skills through on-the-job training.

Training Curriculum Design, Development, and Scheduling

If you know in advance the supply of PMs and BAs in each of the cells and the demand for them based on the projects proposed for those same cells for the portfolio, you can define a training curriculum. The objective of the training program is to have a sufficient supply of PMs and BAs ready to meet the expected project staffing needs. One of my clients had a quite sophisticated process for forecasting the number of professionals of a given position type over time. A career and professional development program existed for each individual in the IT department, and the client could forecast over time how an individual would progress through the plan. This information was used to schedule instructor-led courses for on-the-job and off-the-job training needs.

One of the objectives of an effective training program is to attain and sustain an alignment between the supply and demand for professionals by position type.

Resource Planning Process

The supply of PMs and BAs versus the demand for them is a problem for every organization. I've had clients whose supply of junior-level PMs and BAs was ample but the demand was for senior-level PMs and BAs. There is no short-term solution other than hiring outside contractors or changing the proposed projects to align with available PM and BA staff resources. Neither solution is acceptable. A better approach is to implement a program to align future staffing resources with forecasted project staffing needs. There is some history of past project needs that can be used to forecast future needs. The organization might be planning changes whose implementation will require a shift in PMs and BAs to align with those changes.

By assessing the skills of every PM or BA, an organization can generate a profile of the current staff with respect to how the current staff populates the PM/BA landscape. That is the supply side of your PM/BA staff. It does not account for availability. Next, from project history and the current proposed projects, you can get a sense of the demand for each cell in the PM/BA landscape. For each cell there will be either a surplus of staff or a deficit. That information will inform your HR, Training Director, Project Management Office (PMO) Director, and portfolio managers where the focus needs to be on bringing a balance between the supply and demand for staff.

Resource Management Process

PM and BA staff resources are fixed at least in the short term. Your objective in designing an effective resource management process is to minimize the number of approved projects that are not adequately staffed and the number of PMs and BAs who are "on the bench" because there was no appropriate project for them. In the absence of a project assignment, these individuals can always engage in activities related to their development program.

Putting It All Together

In this chapter I have defined the foundation for resource scheduling, professional development, and a resource-constrained project portfolio management process. The foundation is robust and easily accommodates specific organizational BA and PM position families. The proficiency metric is simple and can be learned quickly by both the individual doing self-assessment and their managers who might also do skills assessment.

CHAPTER 5

The Project Landscape

The majority of current projects are complex and uncertain. The simple projects are all done. No one would argue that. To manage today's complex and uncertain projects effectively requires approaches that are a function not only of project type but also of duration, cost, criticality, business environment, market conditions, client maturity, number of affected departments, and other factors. A one-size-fits-all approach to managing complex and uncertain projects just won't work. A one-size-fits-all approach may never have worked, but we have just come to that realization in the past 20 years. Scrum and Rational Unified Process (RUP) are but two of more than a dozen possible additions to your arsenal but not all complex and uncertain projects are candidates for a Scrum or RUP approach.

If every project is unique, and I think there is universal agreement on that, then why wouldn't one expect the management of these unique projects also to be unique?

For over 20 years I have maintained that the best-fit project management approach is a function of at least the factors listed above. The best-fit project management approach is different for every project because every project is different. We need an objective and repeatable way to tell one type of project from another and then align the best-fit project management approach based on the project type. Therefore, the first order of business is to establish a classification of project types and within that classification scheme discuss the implications to management and staffing approaches for these different types of projects. The discussion of whether the project team needs a PM, BA, or a PM/BA can follow from that discussion. That is one of the reasons I wrote this book. This chapter draws on the project landscape defined in my earlier book (*Effective Project Management: Traditional, Agile, Extreme*, 5th ed. [Hoboken, NJ: John

Wiley & Sons, 2009]) and applies it to management staffing decisions with PMs and BAs.

> *A one-size-fits-all approach to managing complex and uncertain projects just won't work. It may never have worked, but we have just come to that realization in the past 20 years.*

Project Landscape

I'm 100% Polish and, among other things, I like to keep things simple. I see myself as a pretty creative problem solver and approach each situation as being unique. I prefer using graphics to explain concepts and processes but my precondition for those graphics is that they have to be intuitive. Their meaning and intent must be obvious and not require further explanation to make sense to the reader. After a few minutes of observation and thought about any one of my graphics, its meaning unfolds and thoughtful discussion can take place. Designing intuitive graphics is a big challenge, but I thrive on challenges.

Several years ago I was trying to answer the question: What is the best way to graphically and intuitively classify the many types of projects? I wasn't looking for a quantitative metric displayed on a complexity/uncertainty graph or a risk/reward bubble chart. These are valuable but would not help me choose a best-fit project management approach. Rather I was looking for a way to qualitatively define a landscape that I could use as the foundation of a model for classifying a project and then, based on that classification, choosing the best-fit project management methodology. Figure 5.1 is the result of my investigation. At the risk of oversimplifying

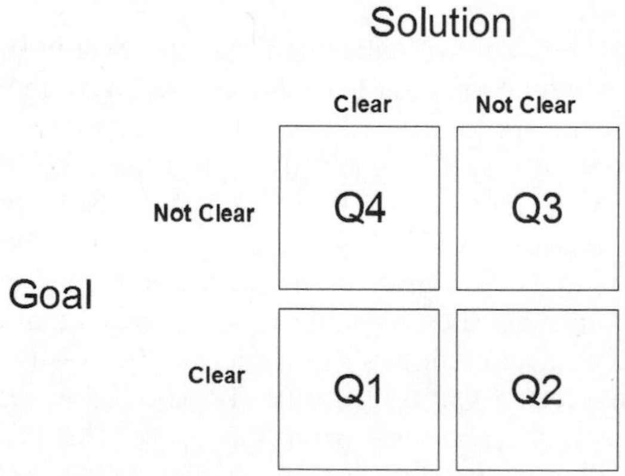

Solution
Clear
Not Clear
Goal
Not Clear
Q4
Q3
Clear
Q1
Q2

FIGURE 5.1 Project Landscape

a complex and uncertain project environment, Figure 5.1 seems to have done the job. Granted, it is only one way to classify project types, but it works and it is intuitive. Two variables define this landscape: the goal and the solution. Every project has a goal, and most goals have a means to achieve them (a solution). Every project has a solution. That solution may be easily attained, or it may never be discovered. Solutions have to be scrutinized to establish the business value they will contribute before they are considered acceptable.

To keep my landscape simple, each dimension (goal and solution) has only two values: Clear or Unclear. The difference between clear and unclear is more qualitative than quantitative. I experimented with three and four values for each dimension (very clear, clear, mostly clear, a little bit unclear, unclear, very unclear or completely unclear), but that only added complexity to each dimension and did not contribute to project classification in any meaningful or useful way. Again, the goal of my landscape is to guide the project team to the best-fit project management approach. I have used Figure 5.1 for over 10 years, and it works.

Every project that ever existed or will exist fits into one and only one of the four quadrants at any point in time.

As the solution develops, the project can change quadrants. For example, projects that start out in Q2 will sooner or later admit of a solution that is clearly defined but not yet realized. This makes it a candidate for Q1. Whether it is moved to Q1 is a complex management decision. Moving a Q2 project to Q1 requires a change of Project Management Life Cycle (PMLC). team members, schedule, and resource requirements. Does the added business value justify the cost of abandonment?

With Figure 5.1 as the foundation, let us look inside that four-quadrant landscape and see what else it is telling us about project type, management approach, and staffing alternatives. My goal is to develop a strategy for staffing a project with PMs and BAs to maximize the expected business value without adversely affecting project risk.

Project Complexity and Uncertainty

Complexity and uncertainty are the driving factors for choosing a best-fit staffing approach. Contemporary projects have become more uncertain, and along with this increased uncertainty is increased complexity and the resulting impact on risk. Uncertainty is the result of changing market conditions that require high-speed and high-change responses to produce a solution in order to be competitive. Complexity is the result of a solution that has eluded detection and will be difficult to find. That imposes a challenge on the PM and BA to be able to respond appropriately. How to assign PM and BA management staff to such projects is central to the choice of best-fit approach.

As you move through the quadrants, the project management processes you use must align with the project. Don't burden yourself and your team with needless planning and documentation if that doesn't increase your chances of finding an acceptable solution. Doing so will just hinder your efforts and detract from the creative process.

> *The idea of enough structure, but not too much, drives agile managers to continually ask the question, "How little structure can I get away with?" Too much structure stifles creativity. Too little structure breeds inefficiency.*
>
> —*Jim Highsmith,* Agile Project Management: Creating Innovative Products (Boston: Addison-Wesley, 2004)

Q1 projects use a plan-driven, process-heavy, and documentation-heavy approach. As you move to Q2 and Q3, project heaviness gives way to lightness. Plan-driven gives way to value-driven. Rigid process gives way to adaptive process, and documentation is largely replaced by tacit knowledge shared among team members. These are some of the characteristics of the many management approaches used for Q2, Q3, and Q4 projects.

Complexity and uncertainty are positively correlated. As projects become more complex, they tend to become more uncertain. Complexity and uncertainty are reflected in the different profiles of projects when it comes to client involvement, requirements, risk, change, team cohesion, communications, and business value. These differences will suggest how best to configure the PM and BA management structure of the project. Let's take a brief look at the project profiles based on the categories just listed.

Client Involvement

I put this at the top of the complexity/uncertainty list because lack of meaningful client involvement is the major reason for project failure (Standish Group, CHAOS 2007 Report). Consider a project where you were most certain of the goal and the solution. Would you bet your firstborn that you had nailed requirements and they would not change? For such a project, you might ask: "Why do I need to have my client involved except for the ceremonial sign-offs at milestone events?" This is a fair question, and ideally you wouldn't need the client's involvement. How about a project at the other extreme, where the goal is a guess and no solution is in sight? In such cases, the meaningful involvement of the client, as a team member, would be indispensable. The PM is not a subject matter expert (SME) and couldn't manage the project without meaningful client involvement.

In Q1 projects, client involvement is usually limited to answering questions as they arise and giving approvals at the appropriate stages of the project life cycle. The assumption with Q1 projects is that requirements are complete and if you build it according to the requirements, they will buy it. It would be accurate to say that client involvement in Q1projects is reactive and passive. But all that changes for projects in the other quadrants. Clients must take a more active role in Q2 projects than in Q1 projects. For Q3 and Q4 projects, meaningful client involvement is essential. The project goes nowhere without that level of commitment from the client. Later in the book we will discuss examples where the BA should be the PM.

The assumption with Q1 projects is that requirements are complete and if you build it according to the requirements, they will buy it.

Finding the solution to a problem is not an individual effort. In Q2 projects, the project team under the co-leadership of the PM and BA is charged with finding the missing parts of the solution. In simple cases, the client will be passively involved, but in complex cases, it is the team under the guidance of the BA that will solve the problem. The willingness of clients to get even passively involved will depend on how you have involved them in the project. They are clearly in a followership role. For Q2, Q3, and Q4 projects, clients or their BA move from a followership role to a collaborative role and even to a leadership role. In your effort to maintain a client-focus and deliver business value, you are dealing with a business problem, not a technology problem. You have to find a business solution. Who is better equipped to help than clients and their BA? After all, you are dealing with their part of the business. Shouldn't they be the best source of help and partnership in finding the solution? You must do whatever it takes to leverage that expertise and insight. Client involvement is so critical that without it, you have no chance of being successful with these complex and uncertain projects.

Establishing ownership by the client of Q2, Q3, and Q4 projects is critical. I often ensure there is that ownership by organizing the project team around co-PMs—one from the developer side and one from the client side. I'd like to suggest that the client-side co-PM is the BA. The co-PMs are equally responsible for the success or failure of the project. That places a vested interest squarely on the shoulders of the client and their BA. This ownership is so important that I have postponed starting client engagements if clients can't send a responsible spokesperson to the planning meeting.

Requirements

As project complexity and uncertainty increases, the likelihood of nailing complete requirements decreases. In a complex software development project, the number of requirements, functionality, features, and dependencies can be staggering. Some will conflict with each other. Some will be redundant. Some will be missing. Many will not become obvious until well into the design, development, and even integration testing tasks. We now have several Q2 models to deal with incomplete requirements. I don't think that should be an obstacle to effective project management. BA involvement throughout the entire project life cycle increases significantly as requirements become less complete and less certain.

The requirements specification document contains more detailed information that might help you decide on the quadrant and management structure of the project. The requirements specification document can be portrayed in a hierarchical requirements breakdown structure (RBS) introduced as Figure 1.3 and reproduced here in Figure 5.2. I have used this

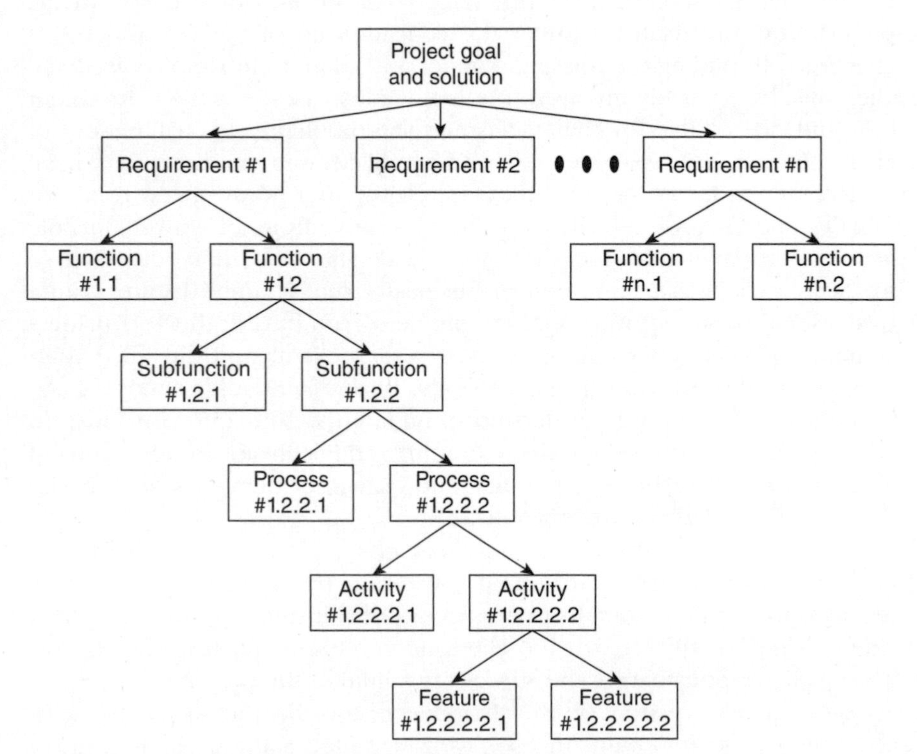

FIGURE 5.2 Requirements Breakdown Structure

successfully with clients for several years. It works because it keeps requirements in a format that is comfortable for the client and it meets my requirement for being an intuitive graphic.

Uncertainty at the requirements level has more impact on your choice of PMLC model than does uncertainty at the functionality level, which has more impact than uncertainty at the feature level. Gauging the completeness of the RBS will always be a subjective assessment. Based on that assessment, a PMLC model is chosen.

The BA should facilitate requirements elicitation, documentation, and management throughout the entire PMLC model and participate with the PM in choosing the PMLC model for the project. No matter how diligent you are, you cannot know what you don't know and you probably don't know that there are parts of the RBS that you don't know. These will become clear as the project progresses and the solution comes into clearer focus.

Risk

Project risk increases as the project falls in Q1, Q2, Q3, and Q4. In Q1 projects, you clearly know the goal and the solution and can build a definitive plan for getting there. The exposure to risks associated with product failure will be low. The focus can then shift to process failure. A list of candidate risk drivers would have been compiled over past similar projects. Their likelihood, impact, and the appropriate mitigations will be known and documented. Like a good athlete, you will have anticipated what might happen and know how to act if it does.

As the project takes on the characteristics of a Q2 project, two forces come into play. First, the PMLC model becomes more flexible and lighter. The process burden lessens as more attention is placed on delivering business value than on compliance with a process. Second, at the same time, product risk increases. Risk increases in relation to the extent to which the solution is not known. On balance, that means more effort should be placed on product risk management as the project moves through Q2 and begins to look more like a Q3 or Q4 project. There will be less experience with these risks because they are specific to the product being developed. In Q3 and Q4 projects, risk is the highest because you are in a research and development environment where project failure is common. There will be numerous product failures because of the highly speculative nature of Q3 and Q4 projects, but these failures are expected. At worst, those failures eliminate one or more paths of investigation and so narrow the range of possible solutions for future projects. These risk management plans require the leadership of the BA more than the PM.

Change

As complexity increases, so does the need to receive and process change requests. Plan-driven project management approaches used with Q1 projects are not designed to respond to change effectively. Change upsets the order of things, and some of the project plan is rendered obsolete and must be redone. Resource schedules are compromised and may have to be renegotiated at some cost. The more that change has to be dealt with, the more time is spent processing and evaluating those changes. That time is forever lost to the project. It should have been spent on value-added work. Instead it was spent processing change requests.

I simulated the impact of change on a simple project. The project was a 12-month project, and it was using a Q1 PMLC. Every month a change request was submitted, and it took three days to process, including analysis of alternatives and project plan revisions. I was surprised to discover that one-third of the total labor time on the project was spent processing change requests!

The BA is critical to the orderly preparation, prioritization, and proposing of change requests. The BA becomes the clearinghouse for all change requests and needs to have a working knowledge of project management to be effective. Once a change request has been submitted, the BA and PM will need to decide when and how to implement the request. This is especially important in the deliverables release schedule.

When it comes to change, Q2 projects are a different story altogether. Any change in the project scope will come about through the normal learning process that takes place in any Q2 project. When clients have the opportunity to examine and experiment with a partial solution, they will invariably come back to the developers with suggestions for other requirements, functionality, and features that should be part of the solution.

In Q3 and Q4 projects, there is yet a further reliance on change to effect a good business-valued product. In fact, Q3 projects require change in order to have any chance at finding a successful solution. Change is the only vehicle that will lead to a solution.

Team Cohesion

In Q1 projects, the successful team doesn't really have to be a team at all. You assemble a group of specialists and assign each to his or her respective tasks at the appropriate times. Period. The plan is sacred, and the plan will guide the team members through their tasks. It will tell them what they need to do, when they need to do it, and how they will know they have finished each task. So the Q1 project plan has to be pretty specific, clear, and complete. Your team members are a group of specialists. Each knows his or her own discipline and is brought to the team to apply that discipline to a set of specific tasks. When they have met their obligation, they often leave the team to return later if needed.

The situation quickly changes for Q2, Q3, and Q4 projects. First of all, there is a gradual shift from a team of specialists to a team of generalists. The team becomes more self-organizing, self-sufficient, and self-directing as the project complexity and uncertainty rises. Q1 teams do not have to be co-located as do teams for any of the other three quadrant projects.

It is highly recommended that Q2, Q3, and Q4 project teams be co-located. Research has shown that co-location adds significantly to the likelihood of successful project completion. Not being co-located creates communication and coordination problems for the PM. One of the first Q2 projects that I managed had a team of 35 professionals scattered across 11 time zones. We were still able to have daily 15-minute team meetings! Despite the communications obstacle, the project was completed successfully, but I have to admit that it added considerably more management overhead than there would have been if the team had been co-located.

Communications

As a project increases in complexity and uncertainty, communication requirements increase and change. When complexity and uncertainty are low, the predominant form of communications is one way (e.g., written). Status reports, change requests, meeting minutes, issues reporting, problem resolution, project plan updates, and other written reports are posted on the project portal and available to any interested parties. As complexity and uncertainty increase, one-way communications have to give way to two-way communications, so written communications give way to meetings and other forums for verbal communication. Distributed team structures give way to co-located team structures to support the change in communications modes. The burden of plan-driven approaches is lightened, and the communications requirements of value-driven approaches take over.

Value-driven communications approaches are the derivatives of meaningful client involvement where discussions generate status updates and plans going forward. Because projects that are high in complexity and uncertainty depend on frequent change, there is a low tolerance of written communications. In these project situations, the preparation, distribution, reading, and responding to written communications is viewed as a heavy burden and just another example of non–value-added work. It is more for historical record keeping than it is for action items. It should be minimized, and the energy should be spent on value-added work.

Business Value

Aren't all projects designed to deliver business value? These projects were commissioned based on an assumption about the business value they would return to the enterprise. This is all true. Q1 management models focus on meeting the plan-driven parameters: time, cost, and scope. When the project was originally proposed, the business climate was such that the proposed solution was the best that could be had. In a static world, that condition would hold. Unfortunately, the business world is not static, and the needs of the client aren't either. The bottom line is this: What will deliver business value is a moving target. Q1 models aren't equipped with the right stuff to assure alignment to the delivery of business value.

Business value increases as you move from Q1to Q2 to Q3 to Q4. At the same time, risk also increases, which means that higher-valued projects are expected in order to be commissionable as you move across the quadrants.

What does this mean? Simply put, whatever PMLC model you adopt for the project, it must be one that allows redirection as business conditions change. The more uncertainty that is present in the development project, the more need there is to be able to redirect the project to take advantage of changing conditions and opportunities. The BA is in the best position to manage that redirection.

Q1 Projects

How could it be any better than to clearly know the goal and the solution? This is the simplest of all possible project situations, but it is also the least likely to occur in today's fast-paced, continuously changing business world. Projects that fall into Q1 are familiar to the organization. Many infrastructure projects will fall into Q1. Perhaps they are similar to projects that have been done several times before. There are no surprises. The client has clearly specified the goal, and the project team has defined how it will reach that goal. Little change is expected. There are different approaches that are in use for such projects, and you will learn how to choose from among them the approach that best fits your project. Such projects also put the team on familiar technology grounds. The hardware, software, and telecommunications environments are familiar to the team. Team members have used them repeatedly and have developed a skilled and competent developer bench to handle such projects.

The limiting factor in these plan-driven approaches is that they are change intolerant. They are focused on delivering according to time and budget constraints and rely more on compliance to plan than on delivering business value. The plan is sacred, and conformance to it is the hallmark of the successful project team.

Because of the times we live in, projects in the Q1 are rapidly becoming dinosaurs. At least the frequency of their application is diminishing rapidly. The simple projects have all been done. The management approaches to Q1 projects are giving way to a whole new collection of approaches that are more client focused and deliver business value rather than strict adherence to a schedule and budget.

In addition to a clearly defined goal and solution, projects that correctly fall into Q1 have several identifying characteristics, as are briefly identified in the next sections.

Low Complexity

Other than the fact that a low-complexity project really is simple, this characteristic often is attributable to the fact that the project rings of familiarity. It may be a straightforward application of established business rules and therefore take advantage of existing designs and coding. Because these projects have been done many times, they often depend on a relatively complete set of templates for their execution. To the developer, it may look like a cut-and-paste exercise. In such cases, integration and testing will be the most challenging phases of the project.

There will be situations where the project is complex but still well defined. These are rare.

Few Scope Change Requests

This is where Q1 management approaches get into trouble. The assumption is that the RBS is relatively complete, and there will be few, if any, scope change requests. Every scope change request requires that these actions be taken:

- Someone needs to decide if the request warrants an analysis by a project team member.
- The Project Manager must assign the request to the appropriate team member.
- The assigned team member conducts the analysis and writes the project impact statement.
- The Project Manager informs the client of the recommendations.
- The Project Manager and client must make a decision as to whether the change will be approved and, if so, how it will be accomplished. If the scope change request is approved, the project scope, cost, schedule, resource requirements, and client acceptance criteria are updated.

All of this takes time away from the team member's schedule commitments. If there are too many scope change requests, you will see the effect they will have on the project schedule. Furthermore, much of the time spent planning the project before the request was made becomes non–value-added time.

The answer to too-frequent scope change requests is some form of management monitoring and control. Those management controls are part of every Q1, Q2, Q3, and Q4 approach but different for every change request.

Well-Understood Technology Infrastructure

A well-understood technology infrastructure is stable and will have been the foundation for many projects in the past. That means the accompanying skills and competencies to work with the technology infrastructure are well grounded in the development teams. If the technology is new or not well understood by the project team, there are alternative strategies for approaching the project. One strategy is for the BA to propose a version 1 solution based on an established technology and at a later date move to a version 2 solution that incorporates the newer technology, which is now better understood.

Low Risk

See Chapter 6 for a complete discussion of risk.

Experienced and Skilled Project Teams

Past projects can be good training grounds for project teams. Team members will have had opportunities to learn or to enhance their skills and competencies through project assignments. These skills and competencies are a critical success factor in all projects. As the characteristics of the deliverables change, so does the profile of the team that can be most effective in developing the deliverables.

Plan-Driven Q1 Projects

Because all of the information that could be known about the project is known and considered stable, the appropriate PMLC model would be the one that gets to the end as quickly as possible. Based on the requirements, desired functionality, and specific features, a complete project plan can be developed. It specifies all of the work that is needed to meet the requirements, the scheduling of that work, and the staff resources needed to deliver the planned work. Q1 projects are clearly plan-driven projects. Their success is measured by compliance and delivery to that plan.

Knowing this, you can use a Q1 management approach for such projects. For example, you can build a complete work breakdown structure (WBS) and from that estimate duration, estimate resource requirements, construct the project schedule, and write the project proposal. This is a nice neat package and seemingly quite straightforward and simple. Oh, that the life of a PM was that simple. But it isn't, and that's where the real challenge comes in.

These are the simplest of projects. Q1 projects have a clear goal, and how to achieve the goal (the solution) is also clear. These are the projects one usually thinks of when one thinks about project management. Would you be surprised to find out that Q1 projects do not occur often? Testimonial data that I have gathered from project, program, and portfolio managers from all over the world suggests that about 20% of all projects are Q1 projects. Think about the projects you have been involved in, and reflect on how often the goal and the solution were both clear. You can recognize these projects because they are projects for which there was little change and the changes that were proposed by the client had at most a minor impact on the project schedule, costs, and deliverables. If the proposed changes were more significant, then you experienced something quite different and probably did not have a Q1 project.

Among the Q1 projects are those that have been done repeatedly. For projects that have been done several times, there are no surprises. The risks are known, and there are well-defined and documented containment and mitigation strategies in place. There is a template WBS in place that is easily modified to reflect the work that had to be done for the specific project at hand. From repeated use the WBS is known to be complete and so a project schedule (a template schedule exists and only needs minor adjustment) has been created. Resource requirements are known as well.

Other Q1 projects are those that barely meet the requirements to be called a project. They are very simple projects by all definitions. Many will involve only one department and no resources outside of those budgeted for the department. Installing a simple change to a business process is an example.

All Q1 projects can be managed using a methodology that I will call Traditional Project Management (TPM). TPM is the foundation of the Project Management Institute (PMI) Project Management Body of Knowledge (PMBOK) and of the models of many professional service firms as well. TPM approaches date from the 1950s. There are two variations of TPM to consider: linear and incremental. The PM and BA management staffing of linear versus incremental TPM projects are different.

Linear TPM Models (Q1)

Linear TPM models are the models that most business managers have in mind when they think about project management. Indeed, they are the simplest models and have been in use longer than any other project management models. Linear TPM models are a good place to start on your journey to learning to become a PM. For software development projects, the waterfall model is the simplest and most familiar example. It dates from the 1950s. The goal and solution are clearly defined. That means a complete WBS can be built and a complete project plan can be developed with all resource requirements known and scheduled. The major assumption that leads to using these models is that since the goal and solution are so well defined, few change requests are expected. Change requests require project team resources to process and, if approved, can have major impact on schedules and resource requirements. That work has already been done once for the initial project plan; repeating much of that effort makes the original effort a wasted one. Just taking the prospect of frequent change requests into account can lead to better choices for the project management approach. The role of the BA can be different for each choice. I've been involved in dozens and dozens of projects, and I have never experienced one that did not have several changes. The world is not a static place, and to expect projects to be static is just plain unrealistic.

Incremental TPM Models (Q1)

Incremental TPM projects have exactly the same characteristics as those projects that use a linear TPM model. That is, the goal and solution are both clearly defined. The only difference in project management approach for these projects is that the client chooses to release the deliverables in stages or increments. In a linear TPM model, all deliverables are released at the end of the project in one release. In an incremental TPM model, the deliverables are released over time. There are two business reasons for using an incremental TPM model:

1. To establish a competitive advantage by getting to market first
2. To begin generating return on investment (ROI) earlier than in linear TPM models

Decomposing the final deliverables into increments is a challenge to effective management of these projects. The role of the BA in these projects is more challenging than in linear TPM models. The BA is the critical component and must collaborate with the PM in decomposing the deliverables because they are all related to the requirements. Decomposing the deliverables requires knowledge of the product/process because the grouping of deliverables across several increments has to produce marketable partial solutions and has to adhere to the dependencies that exist between groups of deliverables. Incremental TPM projects are best staffed by both a PM and a BA working together to decide what will be included in each increment. To do this effectively, the PM needs only minimal skills in business analysis, and the BA needs only minimal skills in project management. The interaction between the PM and the BA is constant throughout the project. There will be changes throughout the project that impact future increments, and increment plans and deliverables will have to be adjusted to accommodate these changes. The BA should be a Manager Level BA/pm with intimate SME knowledge of the business processes affected by the project and of how best to decompose and implement the deliverables. Similarly, the PM should be a Manager Level PM/ba. Each increment is dependent on all previous increments from a management and delivery perspective. An unexpected schedule or resource change in an early increment can have major impacts on all future increments. Those impacts can be significant and will require the undivided attention of the PM.

Example TPM Projects

- Revising the order entry system to accommodate credit card payments over the Internet
- Installing a computer network in a field sales office
- Updating your department or organization to the latest version of a commercial off-the-shelf software application

TPM Team Organization

The simplest of TPM projects, because of the clarity of their goal and solution, can be managed effectively by a BA with minimal project management skills. Here the BA will simultaneously be the PM, but the simplicity of the project does not put an undue burden on the BA. Conversely, a PM with minimal business analysis skills can also effectively manage many of these projects. In this case, the PM will also perform the role of the BA. For the majority of TPM projects, both a BA and a PM will be assigned. Their major interface will be requirements hand-off, which should be fairly straightforward. Refer to Table 1.2 and the discussion after it for an explanation of the relationship between the BA and the PM on TPM projects.

Organizations with a shortage of Manager Level BAs or PMs should give serious consideration to using PMs with business analysis skills or BAs with project management skills to function concurrently in both roles. As the TPM project increases in complexity, duration, cost, or criticality, the decision on how best to staff the management of the project from a PM and BA standpoint will change. The supply of PMs and BAs versus the demand for such professionals and awareness of the resource constraints on the project portfolio are important inputs to project staffing decisions. Effective management staffing of a project should take the entire portfolio into consideration. Those decisions are made with a view toward maximizing the business value of the project portfolio.

Q2 Projects

Q2 projects have a clear goal but how to achieve the goal (the solution) is not clear. These types of projects dominate the landscape. Testimonial data that I have gathered from all over the world suggests that fully 70% of all projects are Q2 projects. Q2 projects have their own project management models, and they are quite different from Q1 models. Unfortunately, too many PMs try to modify Q1 approaches to accommodate Q2 projects. That thinking is a big mistake and a big contributor to the high project failure rate. The range of complexity and uncertainty that characterizes Q2 projects is wide and deep. Q2 projects dwarf Q1 projects with their challenges to the PM and especially to the BA. The least complex and uncertain Q2 projects will require one type of management staffing; those at the highest levels of complexity and uncertainty will require a different type of management staffing. The solution will have different levels of clarity, and it is these different levels of clarity that further suggest specific Q2 project management models. Q2 projects can be managed using methodologies that I will classify as Agile Project Management (APM).

There are various iterative and adaptive approaches to managing APM projects that can be used when the goal is clearly defined but how to reach the goal—the solution—is not. Imagine a continuum of projects that range from situations where almost all of the solution is clearly and completely defined to situations where very little of the solution is clearly and completely defined. This is the range of projects that occupy Q2. As you give some thought to where your projects fall in this quadrant, consider the possibility that many, if not most, of your projects are really Q2 projects. If that is the case, shouldn't you also be considering using an approach to managing these projects that accommodates the goal and solution characteristics of the project rather than trying to force-fit it to some other approach that was designed for projects with much different characteristics?

I contend that the adaptive and iterative class of Q2 projects is continuously growing. I make it a practice at all rubber-chicken dinner presentations to ask about the frequency with which the attendees encounter Q2 projects. With very small variances in their responses, they say that at least 70% of all their projects are Q2 projects, 20% are Q1 projects, and the remaining 10% is split between Q3 and Q4 projects. Unfortunately, many project managers try to apply Q1 approaches to Q2 projects and meet with very little success. The results have ranged from mediocre success to outright failure. Q2 projects present a different challenge and need a different approach. For years I have advocated that the approach to the project must be driven by the characteristics of the project. To reverse the order is to court disaster. I find it puzzling that we define a project as a unique experience that has never happened before and will never happen again under the same set of circumstances, but we make no assertion that the appropriate project management approach for these unique projects will also be unique. I would say that the project management approach is unique up to a point. Its uniqueness is constrained to using a set of validated and certified tools, templates, and processes. To not establish such a boundary on how you can manage a project would be chaotic. Plus the organization could never be a learning organization when it comes to project management processes and practices.

As the solution moves from those that are clearly specified toward those that are not clearly specified, you move through a number of situations that require different handling. For example, suppose only some minor aspects of the solution are not known; say the background and font color for the login screens. How would you proceed? An approach that includes as much of the solution as is known at the time should work quite well. That approach would allow the client to examine, in the sense of a production prototype, what is in the solution in an attempt to discover what is not in the solution but should be. At the extreme, when very little is known about the solution, projects have higher risk than those where a larger part of the solution is known. A solution is needed, and it is important that a solution be found. How would you proceed? What is needed is an approach that is designed to learn and discover the solution. Somehow that approach must start with what is known and reach out to what is not known.

There are several approaches to Q2 projects. Q1 approaches cannot be used to build a complete WBS for Q2 projects without guessing. Because guessing is unacceptable in good project planning, you have to choose an approach designed to work in the absence of the complete WBS. All Q2 approaches are structured so that you will be able to learn and discover the missing parts of the solution. As these missing parts are discovered, they are integrated into the solution. There are two distinct PMLC models for use in Q2 projects: iterative PMLC models and adaptive PMLC models. The choice of which model to use depends somewhat on the initial degree of uncertainty you have about the solution.

Projects that correctly use a Q2 approach have several defining characteristics, as briefly identified in the next sections.

A Critical Problem without a Known Solution

Critical problems without known solutions are projects that must be done. You have no choice. Because there is no known solution, a Q1 approach, which requires a complete RBS and WBS, will not work. You must use a Q2 approach. Despite the realities, it amuses me how many PMs try to use a hammer when a screwdriver is needed. (Maybe some of them only have hammers.) The only approaches that make sense are those that enable you to discover an acceptable solution by doing the project. These projects fly in the face of all of the traditional practices of project management. Some of the functions and features of the solution may be known, but there is not enough business value in the known partial solution for it to be implemented.

Previously Untapped Business Opportunity

In previously untapped business opportunities, the company is losing out on a business opportunity and must find a way to take advantage of it through a new or revamped product or service offering. The question is: What is that business opportunity, and how can you take advantage of it? Here very little of the solution is known.

Q2 Project Are Critical to the Organization

You should have guessed by now that a Q2 project can be very high risk. If previous attempts to solve the problem have failed, it means the problem is complex and there may not be an acceptable solution to it. The organization will just have to live with that reality and make the best of it. Projects to find that elusive solution might work better if they are focused on parts of the problem or if approached as process improvement projects.

Meaningful Client Involvement Is Essential

The solution will be discovered only if the client and the development team collaborate meaningfully in an open and honest environment. For the client this means fully participating with the project team and being willing to learn how to be a client in an agile world. For the development team, this means a willingness to learn about the client's business and how to communicate in the client's language. For the project manager, this means preparing both the client team and the development team to work together in an open and collaborative environment. It also means that the project manager will have to share responsibility and leadership with a client manager.

Q2 Projects Use Small Co-Located Teams

If the project requires a team of more than 30 professionals, you probably should divide it into several smaller projects with more limited scopes. As a rule, Q2 approaches do not scale well. To manage a 30+ project team, partition it into smaller teams, with each of these teams being responsible for part of the scope. Set up a temporary program office to manage and coordinate the work of the smaller project teams.

There are two very different Q2 variations to consider: Iterative and Adaptive.

Iterative APM Models (Q2)

Iterative APM projects are used in the simpler of the two types of Q2 projects. Iterative APM models are appropriate for projects whose solution is nearly complete. There may be several known alternatives for the few missing pieces of the solution. The project will have to determine which alternatives best meet client requirements. This will be done through a number of iterations until the complete solution comes into focus. Prototyping is a good example of an iterative APM model. As long as there is some reasonable idea as to what the solution will look like, the developer can build on that and eventually get a complete picture of the solution. While prototyping probably can be used on the most complex of Q2 projects, there are better choices. The evolutionary development waterfall model and the feature-driven development model are examples of iterative APM models familiar to software developers. The RUP model is classified by many developers as an iterative APM model.

Adaptive APM Models (Q2)

At the other end of the Q2 project landscape are those projects whose solution is almost totally unknown. These projects are the most complex and uncertain of the Q2 projects. They may have been attempted several times before with little or no success. These will typically be very challenging and high risk because the search for an acceptable solution is striking out into the great unknown, and no one can predict the outcome. Perhaps there is no solution that meets the goal, and that is the reason all previous attempts have failed. Or maybe there is a solution but it only partially meets the goal and implementing it didn't make sense. Obviously these projects will be mission-critical projects, and some solution must be found. No approach other than adaptive will work on them. You cannot force fit these projects into iterative APM or Q1 models. I have seen numerous cases of such attempts, and they have all failed. If these projects are successful in finding a solution that meets the goal, there will be a high business value return to the organization. If that were not the case, such projects would not be high-priority ones. But the risk is high and stacked against success.

Several Q2 models work across the spectrum of agile software development projects. Dynamic Systems Development Method (DSDM), Scrum, and Adaptive Software Development (ASD) are three, but there are several others. Adaptive Project Framework (APF) is an agile approach that I developed for projects that do not involve software development, but it also can be used on software development projects. APF grew out of two independent client engagements, one involving kiosk design and the other involving process redesign.

APF's real strength is with projects where almost none of the solution is known at the outset. The solution is discovered through iteration, where each iteration has both probative and integrative swim lanes. The probative swim lanes test the impact of hypothetical additions to the then-current solution. The successful additions are prioritized for implementation in an integrative swim lane. Integrative swim lanes incorporate discovered solution parts. For a complete discussion of APF, see my book *Adaptive Project Framework: Managing Complexity in the Face of Uncertainty* (Boston: Addison-Wesley, 2010).

Example APM Projects

This example predates any APM model but is instructive nonetheless. In the 1961 State of the Union message President John F. Kennedy made this statement: "I believe that this nation should commit itself to achieving this goal before this decade is out, of landing a man on the moon and returning him safely to the earth." I was fortunate to be involved in that project, and I know that none of the engineers I worked with had any idea how that goal was to be reached. So the solution was almost totally unknown in 1963. This would have been an example of a highly complex and uncertain APM Project, if APM existed. My company was responsible for figuring out if a spacecraft could safely be powered by a nuclear reactor. The radioactive environment created by the nuclear reactor would reduce the onboard computer and the crew to dust, so some creative solution (if one could be found) would be needed. Since there was no way to test solutions in the real world, some form of simulation model would have to be developed. I was one of the few systems consultants with the mathematical skills to create such a model. Through the simulation model I was to search for a solution using a nuclear reactor, if such a solution existed. If a solution were found, it would have to be practical and feasible from a cost perspective.

I was able to find a solution that used a nuclear reactor as the power source. At the time paraffin wax was the best insulator against radioactive environments. The simulation model tested several alternative configurations and feasible spacecraft designs suggested by NASA and determined that if the paraffin wax shield were one mile thick, it would protect the computer and crew from the adverse effects of the radiation. I haven't seen any one-mile-wide spacecraft yet so I suppose my solution was not feasible.

A more relevant and timely example is to design and develop a nursing staff scheduling system. The functionality that an acceptable solution had to deliver was well defined but how to achieve that functionality was not. So the goal is clear but the solution was not. Finding an acceptable solution, if one could be found, would require very complex mathematical modeling. To date there is no closed-form mathematical solution to this problem. Solutions using heuristic algorithms or decision support systems do exist. There is no way of knowing whether these are optimal, but they are acceptable.

Another example is in the process improvement area. Virtually every process can be improved. Take a simple example: A pizza chain guarantees 45-minute order entry to order delivery for in-home delivery service. It has lost a significant part of their sales because of competitors who offer 30-minute service. The chain is not competitive and has to reduce its service time to 30 minutes in order to stay in the home delivery business. The service improvement from 45 to 30 minutes is significant and may not be achievable, but no one really knows at the outset if the goal is achievable.

All process improvement projects are of this type. A goal in performance is established, and a team is charged to improve the supporting processes to reach the goal. From the standpoint of PMs and BAs, who should staff such projects? How would you handle such situations?

APM Team Organization

By now you should see that staffing PMs and BAs to an APM project is not at all routine. Availability is not a skill! These projects need as much SME presence on the team as is possible. The more complex the project, the more need there is for SMEs. They could be PMs who are specialists and BAs who are both specialists and generalists. The teams should be small (8 to 10 members). If acceptable solutions are beginning to emerge, team size may be increased to focus more efforts on the likely solution.

As you consider PM and BA staffing and role responsibilities, keep in mind that these are the most challenging of projects your organization will ever encounter. Your project portfolio must anticipate these projects and be prepared to staff them using the very best strategies possible. Second best is not an alternative, and neither is project failure. That means you need to put career and professional development programs in place to prepare for the day when you will need many of these professionals to lead and staff your APM projects.

Q3 Projects

Q3 projects do not have a clear goal or solution. The only projects that occupy this part of the project landscape are research and development (R&D) projects. Extreme Project Management (xPM) models are used in Q3 projects. xPM extends to the remotest boundaries of the project landscape. xPM projects are ones whose goal and solution cannot be clearly defined. For example, R&D projects are xPM projects. What little planning is done is done just in time, and the project proceeds through several phases until it converges on an acceptable goal and solution. Clearly the PMLC for an xPM project requires maximum flexibility for the project team, in contrast to the PMLC for a Q1 project, which requires adherence to a defined process. If there isn't any prospect of convergence in the Q3 project,, the client may pull the plug and cancel the project at any time and look for alternative approaches.

If goal clarity is not possible at the beginning of the project, the situation is much like a pure R&D project. Now how would you proceed? In this case, you use an approach that clarifies the goal and contributes to the solution at the same time. The approach must embrace a number of concurrent probative swim lanes. Concurrent probative swim lanes might be the most likely ones that can accomplish goal clarification and the solution set at the same time. Depending on time, budget, and staff resources, these probes might be pursued sequentially or concurrently. Alternatively, the probes might eliminate and narrow the domain of feasible goal/solution pairs. Clearly, Q3 projects are an entirely different class of projects and require a different approach to be successful.

The goal is often not much more than a guess at a desired end state with the hope that a solution to achieve it can be found. In most cases, some modified version of the goal statement is achieved. In other words, the goal and the solution converge on something that one hopes has business value.

In addition to a goal and solution where neither is clearly defined, projects that fall into Q3 have several identifying characteristics, as briefly identified in the next sections.

The Q3 Project Is a Research and Development Project

The goal of an R&D project may be little more than a guess at a desired end state. Whether it is achievable and to what specificity are questions to be answered by the project. In this type of Q3 project, you are looking for some future state that will be realized through some enabling solution. Because you don't know what the solution is, you cannot possibly know what the goal is. The hope is that the goal can be achieved with a solution and that the two together make business sense.

The Q3 Project Is Very High Risk

Any journey into the great unknown is fraught with risk. In the case of a Q3 project, it is the risk of project failure, and that is very high. For example, the direction chosen in the search for the solution may be the wrong direction entirely and can only result in failure. If the project management process can detect that early, it will save money and time.

Failure is difficult to define in a Q3 project. For example, the project may not solve the original problem, but it may deliver a product that has uses elsewhere. The Post-it Note Project is one such example. Nearly seven years after the project to develop an adhesive with certain temporary sticking properties failed (it was a Q3 project), an engineer discovered an application that resulted in the Post-it Note product (that was a Q4 project).

Example Q3 Projects

My favorite example is the project to find a cure for the common cold. While the goal is admirable, it is certainly not clear. What does "cure" mean? There are at least three possibilities for a definition of cure:

1. Prior to birth, inject a DNA-altering substance into the fetus that forever prevents the onset of a cold.
2. Adapt the juice extract from a tree that grows only in the most mountainous regions of Asia so that a daily 8-ounce dose of the liquid prevents a cold from ever occurring.
3. Once you contract a cold, take this pill and in 24 hours all signs that you ever contracted the cold will be gone.

Any one of these might be considered a cure, depending on your definition of "cure." In the absence of any further clarification, the project commences. There might be a deliverable from this project that offers some type of cure, but no one knows what that might be and whether the cure will have any business value. I am told by one of my clients, a large drug researcher, that cures for several diseases are available but have not been produced because they are not commercially viable. One wonders what value has been placed on human life.

Q3 Team Organization

First of all, the team must be heavily staffed by SMEs. Depending on the nature of the business unit, product, or process under study, some of these SMEs might also be BAs. A Q3 project is usually best managed by an approach that is the least intrusive as management will permit. The effort is creative and thrives on discovery. Too much project management structure is viewed as a nuisance and a waste of time and effort. That means that strong PM skills are not necessary. Of more value are the BA skills that can be assigned to the project. These projects are a good example of projects where a BA can also have PM responsibilities.

Q4 Projects

Q4 projects have a clear solution but the goal is not clear. This sounds like a solution out looking for a problem and may seem like nonsense at first, but that is not the case. Like Q3 projects, these are also R&D projects but with a twist.

I call these projects emertxe projects and they use an approach that I call Emertxe Project Management (MPx). If you haven't already guessed it, emertxe is "extreme" spelled backward. I chose that name because Q4 projects will use the same management approach as Q3 projects except the time scale is reversed. Let me explain. In a Q3 project, there is a goal that will be clarified only when a best-fit solution is found. That best-fit solution will meet some goal that is a special case of and probably a more limited version of the original goal. In a Q4 project, you have a solution and are asked to define the goal that it meets. So there was no goal statement at the outset, but the solution will define the goal. Just as in a Q3 project, the question now becomes: Do the goal and solution deliver acceptable business value? If they do, you have succeeded. The major difference between a Q3 and Q4 project is that in a Q3 project, both the goal and the solution converge on a goal and the solution that supports it. In a Q4 project, the solution is fixed and you have to find the goal that it supports. In both Q3 and Q4 projects, the resulting goal and its solution must pass the test of being acceptable from a business value standpoint.

In addition to a goal that is not clearly defined and a solution that is clearly defined, projects that correctly fall into Q4 have several identifying characteristics, as briefly identified in the next sections.

New Technology without a Known Application

Think of technology in the broadest terms. For example, it could be an extract from some tree or plant that seems to have medicinal value to the indigenous people who use it. Does this extract actually have any medicinal properties, and what are they? Can it be used to develop a commercially viable product? You have a solution (the extract). What problem does it solve (the goal)?

Solution Looking for a Problem to Solve

Commercial off-the-shelf application software provides several examples of solutions looking for problems to solve. For example, say, a new Human Resource Management System (HRMS) has just been introduced to the commercial software market by a major software manufacturer. Your project is to evaluate it for possible fit in the new HRMS design that has just been approved by your senior management team. Among all MPx projects, this example is the simplest case. You already know the application area. What you need to find out is the degree of fit and business value. At the other extreme would be to have something whose application is not known. A juice taken from the root of some strange Amazon tree would be an example of a more complex situation. The project is to find an application for the juice that has sufficient business value.

Example Q4 Project

Early in the development of radio-frequency identification (RFID), several companies were investigating applications of the technology in their warehousing and distribution systems. Wal-Mart was a client of mine at the time, and it was one of those companies. The solution was RFID and its technical specification and performance metrics were known. The goal was to find some commercial application of the solution, and it was unknown. Like Q3 projects, maybe there was no achievable goal that delivered acceptable business value. In Wal-Mart's case, early RFID products did not meet its performance requirements, so the Q4 project did not find a viable business application. Interestingly, Wal-Mart was the 800-pound gorilla in the room, and it quickly influenced further RFID research and development (a Q3 project, by the way) to where it is now an accepted and viable business application.

Q4 Team Organization

Q4 project teams are staffed the same as Q3 project teams. The focus is on SMEs and BAs. Project management should avoid being intrusive but instead be supportive of the creative and discovery mode. The manager of a Q3 or Q4 project should be a BA/PM or BA/pm. And that manager is responsible for both the business analysis and project management roles of the project.

Project Landscape with Project Management Models

Figure 5.3 summarizes the preceding discussion.

As I travel around the planet speaking to BAs and PMs at conferences and workshops, I always ask them what percentage of their projects fall in each of the four quadrants. I've asked that question to over 5,000 BAs and PMs in the United States, Canada, Trinidad, England, Germany, Switzerland, Czechoslovakia, China, and India. The results are remarkably consistent:

- TPM (linear and incremental approaches) 20%
- APM (iterative and adaptive approaches) 70%
- xPM/MPx (extreme approaches) 10%

I suspect that a major contributor to project failure is the force-fitting of a linear or incremental approach to a project when an iterative, adaptive, or even extreme approach should have been used.

We say that every project is unique, that it has never happened before and will never happen again under the same set of circumstances. It would be naive then to think that one project management approach would work for every project. I don't know anyone with such clairvoyant skills. We

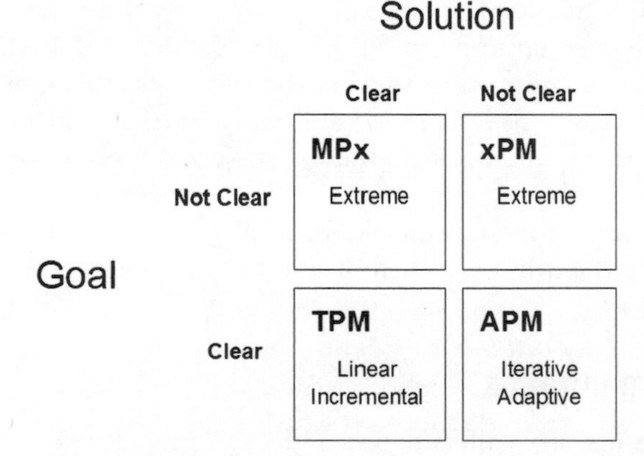

FIGURE 5.3 Project Management Models Loaded into the Project Landscape

have already noted how goal and solution clarity and completeness of requirements drive the choice of development model and project management approach, but several other project characteristics should be considered. I have had occasion to consider risk, cost, duration, complexity, market stability, business value, technology used, business climate, degree to which you expect to have meaningful customer involvement, number of departments affected, the organization's environment, and team skills and proficiencies.

Putting It All Together

I believe in and have always presented a one-stop-shopping experience to my clients. It is critical to project success that a strong sense of teamwork be created between the client and the team and the project manager and her team. The PM/BA professional is better equipped to do that than if a BA and PM were separately involved. The BA, PM, and client structure requires three communications links all working in harmony while the BA/PM requires only one. With people-to-people communications being the major reason for project failure, we need to give serious thought to creating the PM/BA professional to manage APM, xPM, and MPx projects. There is much to discuss about the preparation and development of the PM/BA and BA/PM. Chapter 6 discusses the management staffing situations. Chapter 7 discusses the career and professional development of these people.

CHAPTER 6

Integrating the Project Manager and Business Analyst into the Landscape

This chapter completes my contribution to the infrastructure and definition of the PM and BA staffing situation. It maps the BA and PM professionals into the project landscape defined in Chapter 5. Strengths, weaknesses, challenges and opportunities are discussed. We are at the very frontiers of defining these two very important professionals and how they should interact for maximum benefit to the organization. I am reminded of a statement made over 40 years ago that has great relevance to our present situation:

> *Advanced technology required the collaboration of diverse professions and organizations, often with ambiguous or highly interdependent jurisdictions. In such situations, many of our highly touted rational management techniques break down; and new non-engineering approaches are necessary for the solution of these "systems" problems.*
>
> —Leonard R. Sayles and Margaret K. Chandler, *Managing Large Systems: The Large-Scale Approach* (New York: Harper & Row, 1971)

We are faced with a staffing situation that has not been sufficiently addressed to date. No one doubts that there is a great deal of overlap between the roles and responsibilities of a PM and a BA as they interact on the same project. The Project Management Body of Knowledge (PMBOK) and Business Analysis Body of Knowledge (BABOK) offer details. Who is responsible for what and when? I wish the answer was clear, but life isn't that simple. I'd like to proceed under the assumption that every PM and BA interaction is unique and deserves an open exchange of ideas between the PM and the BA as to how they should work together. There is no formula to be shared. Rather my goal is to provide the thinking and reasoning that goes behind any decision about staffing a project with these professionals. In many projects they will be functioning as co-PMs, and that requires them to decide how best to complement one another's strengths and weaknesses and work together. The goal of delivering maximum business value to the client should remain uppermost in whatever decision they make.

Strengths, Weaknesses, Challenges, and Opportunities of PM and BA Integration

When I use the term "integration," I mean the merging of the roles and responsibilities of the PM and the BA assigned to a project for maximum benefit to the client and the organization. This merging cannot take place until the PM and BA participants on the project have considered the project situation and made the best decision regarding their cooperation and collaboration. While their decision may have been the best choice at that time, they should have a process for periodically reviewing the conditions that led to that decision and reaffirming their earlier decision or changing it to suit current conditions.

Strengths of the PM and BA Integration

There are projects whose successful completion has eluded the organization even in the face of repeated attempts. In some cases finding an acceptable solution to correct an anomaly in a product or a performance issue in a business process is critical to the success or even survival of the organization. These projects are the most challenging a PM will ever encounter, and the best staffing solution available must be put in place. That means having personnel on the team who have expertise in both project management and business analysis. Having these dual skill profiles in no way implies that these professionals will have a dual project responsibility for both project management and business analysis. They might have, but that is the topic of another discussion.

What does make sense to me, at least for Agile Project Management (APM) projects, is PM and BA integration into a co-PM role. The more complexity and uncertainty associated with the project, the greater the need for this co-PM role. Both the PM and the BA equally share responsibility for the outcome of the project. The recommended pairing for the most complex and uncertain projects (those projects that use adaptive project management life cycles [PMLCs]) is that both the PM and the BA occupy Senior Manager Level positions in the PM/BA and BA/PM sectors. For Q2 projects of lesser complexity and uncertainty (those projects that use Iterative PMLCs), it is recommended that the PM and the BA occupy Manager Level positions in the PM/ba and BA/pm sectors as a minimum. It is hoped that this management balance will eliminate any suggestion of a pecking order between the two professionals.

Weaknesses of the PM and BA Integration

If the project is progressing well under the co-PM staffing model, the management problems will be minimal and the co-PMs will be of one mind. Nothing breeds satisfaction and cooperation more than success. Problems arise when the project is not progressing as hoped. Assuming that the co-PMs are really equally responsible, they must agree on all decisions and actions. If one of the pair digs in, the outcome could be disaster for the project. Could this be a place where a decision made by consensus would be the best way to go? I would be cautious about using that approach, given that consensus decisions often are not good decisions. Rather I would favor some other type of tiebreaker. For example, defer to the judgment of the BA/PM on matters related to the product or business process and to the PM/BA on matters related to the project management process. Alternatively, the decision might be based on what is best in terms of delivering acceptable business value. I am assuming an open and good faith effort by both co-PMs.

Challenges to PM and BA Integration

In our case we are wrestling with the management structure for projects of high complexity and uncertainty that involve disciplines that have highly interdependent jurisdictions. The criterion for solving these management structure problems that must be kept uppermost in our minds is that the business value we deliver to our client must always prevail.

The question to be answered about having separate BAs and PMs or a single BA/PM assigned to the same project is certainly based on the characteristics of the project, staff availability, and the internal/external environment in which the project is to be undertaken. To date, no clear answer has been forthcoming. Figure 6.1 gives us some structure to discussing those challenges.

As shown in the figure, there are five environments that impact the decisions we make about the best way to staff a project: the market, the organization, the client, available staff, and the project itself. The organization is a participant within a market, and that market has certain characteristics that will affect the organization and its projects. Within the

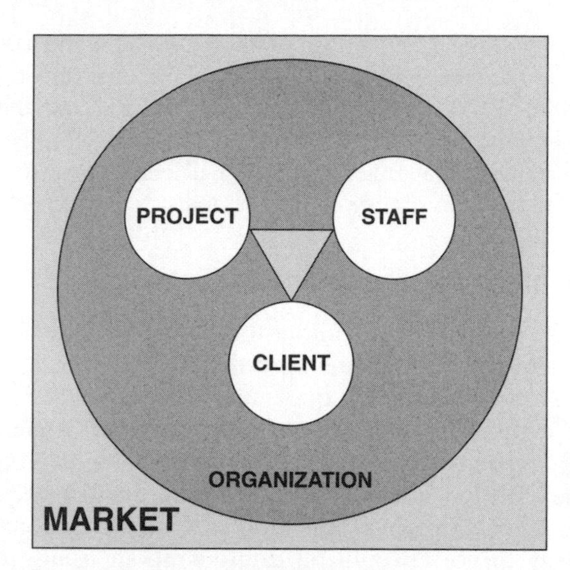

FIGURE 6.1 Internal/External Project Environment

organization is a client group that is requesting a project. That client group can be a single business unit or several business units participating in a single project. The staff members who are available to staff the project will have a skill/proficiency profile that is in alignment with the needs of the project or have some skill gaps that may be critical to the success of the project. And, finally, there is the project, with a whole host of variables and constraints that must be accounted for in whatever staffing decisions are made. Each one can have an impact on project staffing at the project and business management position levels. Let's expand our look at each of these five environments and how management level staffing might be impacted.

PROJECT ENVIRONMENT The project environment is very complex and dynamic. It changes every day, and a staffing decision made today might be different if done tomorrow. Because most projects are complex and uncertain, it is not always clear which decision alternative is best. Decisions about these projects often have to be made with incomplete and frequently changing information. Some of the project environment factors that affect staffing are discussed next.

Most projects that involve a BA are either process/product design or process/product improvement projects. That means the project team will require one or more BA subject matter experts (SMEs) at some depth and some level of involvement. BAs at the staff level will work under the direct supervision of the project manager—a PM/ba. This should work well for design projects. Improvement projects have different BA and subject matter expert (SME) staffing needs. It is probably safe to assume that improvements in the product or process will not be easily found or implemented. The best thinking available at the time designed the product or process under consideration. The world has changed in some way, and improvements are now called for. As the complexity and criticality of these APM projects increase, the skill and proficiency level of the management team assigned to them must also increase.

Complexity becomes a factor in all APM projects. The fact that a complete and acceptable solution is not known raises the possibility that one does not exist or that one might exist but despite best efforts is not found. At the outset, no one ever knows if an acceptable solution does or does not exist. The best we can do is evaluate the known situation and decide if commissioning the project to find a solution is the correct decision. The outcome can be either that an acceptable solution is found and everyone is satisfied or that no acceptable solution is found and the business is no better off for having expended the effort. The project team and the client learned where not to look for a solution but not where to look.

Risk The requirement for Q1 projects is that their environment is known and predictable. There are no surprises. All that could happen to put the project at risk has occurred in the past, and there are well-tested and well-used mitigation strategies that can be used. Experience has rooted out all of the mistakes that could be made. Clients are confident that they have done a great job identifying requirements, functions, and features, and they are not likely to change. The project manager has anticipated and prepared for likely events (not including acts of nature and other unavoidable occurrences). There will be few unanticipated risks in Q1 projects. That doesn't mean you can skip the risk management process in these projects. That will never be the case, regardless of the quadrant the project occupies. However, the intensity, analysis, monitoring, and mitigation strategies will be different in each quadrant.

Risk increases significantly as the project category moves from Q1 to Q2 to Q3 and finally to Q4 projects. Similarly, BA involvement as well as skill and proficiency profile should also increase. But the BA must be an SME for the client because that is what these projects call for. BA generalists can be of help with the process in general, but the BA/SME will be indispensible when it comes to content and understanding of a specific business process.

Effective risk management is critical to projects characterized by complexity and uncertainty. Some risks do not involve the client, but that doesn't absolve the BA, who is the co-PM, from participating in the risk management life cycle. The BA becomes the eyes and ears for the client when it comes to business unit and market conditions that might impact project outcomes.

A risk is some future event that may or may not occur, and you have no control over when it occurs. If it does occur, the project will suffer some loss. The PM, BA, SMEs, or client may be able to take some action that reduces the likelihood of the event occurring and the loss that results from the event having occurred. For example, if the client is change resistant, the BA may be able to actively promote the benefits to the change, thereby reducing any resistance from the client during implementation.

Risk Identification There are four risk categories: technical, project management, organizational, and external. Table 6.1 is the template I use for identifying the operative risks for a project. PM, BA, and SME involvement in the identification and assessment process is recommended. For each identified risk, the impact on the five variables can be assessed.

Risk Assessment There are two components to consider in assessing risk. The first is the probability that the event will occur. The second is the loss that results from the event occurring. For example, if the new business process does not integrate with other dependent processes, then the implementation will be delayed, costing the client $500,000. Let's say that the probability of the event is 0.5. So the actual dollar impact is $500,000 times 0.5, or $250,000. Obviously the risks that have high impact and high probability will draw attention, and some mitigation plan should be put in place. The cost of the mitigation plan needs to be compared to the expected loss. You wouldn't want to implement a $1,000 solution to a $100 problem.

Risk Mitigation A mitigation plan identifies actions the PM or BA can take that will reduce either the likelihood of the event occurring or the loss that would result if it did occur. For any risk there are five possible mitigation plans:

TABLE 6.1 Risk Identification Template

Risks by Risk Categories	Scope	Time	Cost	Quality	Resources
Technical					
Project Management					
Organizational					
External					

1. **Accept the risk** and do nothing because the action is more expensive than the expected loss.
2. **Avoid the risk** by deleting the parts of the project that expose it to the risk.
3. **Develop contingency** plans that identify actions you will take if the event occurs.
4. **Identify mitigating actions** you will take to reduce the impact and/or likelihood of the occurrence of the event.
5. **Transfer** the risk to an outside entity (outsourcing or insurance policy).

The BA has a better perspective on identifying and developing mitigation plans that protect the client's products and processes than does the PM.

Risk Monitoring and Control Someone on the project team is assigned to monitor the status of each risk. One person may have monitoring responsibility for all risks, or each risk might have its own person to monitor status. The PM, BA, or SME might be the best choice, depending on the type of risk. The point I am making is that each risk is owned by the best-fit person. This is a change from current practices, where one person owns the monitoring responsibility for all risks and reports status changes at each team meeting. As a general rule, technical risks are owned by the PM or the SME; project management risks are owned by the PM; organizational and external risks are owned by the BA.

Duration The longer the duration of the project, the greater the likelihood that it will not complete successfully. The frequency of client change requests increases as well. Many of these changes are the result of market changes rather than internal business changes. The longer the project, the more likely there will be staffing changes along the way. Management roles, responsibilities, and staffing will likely change over the course of the project. Project scope often will be the cause of these changes. The project structure could change to a program structure—in effect, decomposing the project into dependent subprojects. In those situations, a new program management structure will have to be put in place. The BA's role will be to decompose the deliverables and deliver them according to some incremental schedule.

Technology Using the latest and greatest technology may not be the best business strategy. Getting to market with a product that uses an established technology is far less risky. Following the release of that product version with an updated version using the more recent technology may be a better long-term strategy. The challenge is to have more senior-level PM and BA staff available for these projects.

CLIENT ENVIRONMENT The latest report from the Standish Group puts client involvement at the top of the list of the reasons why projects fail. That is the first time client involvement has been at the top of its list. The message is clear. We all have to do whatever possible to create the most open and honest relationship with our clients as is possible. In most cases the client does not feel like an equal member of the project team. Both the PM and the BA can play an integral part in diffusing that feeling.

Client involvement is not always predictable and can be very elusive at times. On some projects client involvement will be exemplary. On other projects the same client will be distant and uninterested. I've even had situations where there was so much client involvement that it became an obstacle to project performance. Clients were taking too much time away from the development team with their invasive behaviors and suggestions for change. I've long felt that the best way to start a project is to hold a kick-off meeting with the client and the BA to establish a clear understanding of the rules of the engagement. If there is any reason to believe that there is some disconnect, a workshop should be held immediately to clarify those rules. The complexity and uncertainty that most projects present to the entire team requires that there is a clear understanding so that everyone is on the same page.

Criticality of the Project to the Business The most critical and highest business valued projects demand the best management structure possible. Wal-Mart offers some good advice. Its 285 client groups within IT ensure that the same cadre of professionals is available for the project team that serves the client. Over time the team members become very familiar with their fellow professionals and their work styles and habits. The teams are very effective. High risk usually accompanies these projects. A PM/BA or BA/PM Senior Manager is my choice for the PM. In a more general setting the BA/PM Senior Manager would be a better choice to manage projects for clients whose meaningful involvement has been problematic in the past.

Change Uncontrolled and frivolous change requests from the client are the bane of most PMs. In a Traditional Project Management (TPM) project, they can completely disrupt all of the effort that went into creating the project plan. Conversely, in an APM or Extreme Project Management (xPM) project, their absence is disastrous. The project will never converge on an acceptable solution in the absence of client change requests. In both cases the change requests occupy team members in analyzing the requests and their impact on the project plan. Schedule, cost, and resource requirements are affected. Too many change requests will seriously detract from time available to be spent on building the deliverables. For projects that will likely generate a steady flow of client change requests, having a BA/pm or BA/PM on the team is indispensable. Their role with respect to change requests includes:

- Working with the client to validate and document the change request.
- Prioritizing the new change request with all other change requests from this client for this project.
- Agreeing with the client on the change request submission strategy.
- Proposing the change request to the PM.

They should be senior enough in the discipline and have enough credibility with clients to have full responsibility for managing the change process on the project. Scope change management will remain the responsibility of the PM, as it should. The BA will be the intermediary between the client and the PM in negotiating and accepting change requests. For highly complex and uncertain APM projects, a BA/PM should be assigned to manage the change process.

Client Originates the Project The BA does not have to sell the business value of the idea for the project. If the client exhibits unbridled enthusiasm for the project, the BA would do well to independently justify the business reason for the project and ensure that the expected business value is not fantasy but can be achieved realistically. The Conditions of Satisfaction (COS) process described in Chapter 1 is a critical part of those deliberations. If BAs originate the project, they had better be prepared to sell the project to clients. Clients can easily be taken outside of their comfort zone by not understanding the project, and their support will be more difficult to gain. The BA will be advocating change in some way, and clients will want to know what's in it for them. Clients will internalize their cost (dollars, labor, change, etc.) and compare that against real benefits in deciding whether to support the project. The clients (not their BAs) are responsible for the return on investment and have to own the project before it makes sense for the BA to go forward proposing the project.

Requirements The popular opinion among thought leaders is that complete requirements cannot be gathered at the beginning of any project with the exception perhaps of simple trivial projects. The major reasons are:

- Don't know how the market and competitors will change.
- Finding a complete and acceptable solution requires iteration.
- The proposed solution may not deliver acceptable business value.

I think it is important that we all understand these reasons and staff the project scoping phase with the idea of minimizing the impact of incomplete requirements on the choice of requirements-gathering approaches, on how to manage requirements changes throughout the project, and on managing project performance relative to achieving acceptable business value from the deliverables.

How does this information help with project staffing decisions? The less clients know about what an acceptable solution will look like, the more talent that must be assigned at the management level. There is an ordering that may help. The next management level staff are recommended as you move from most complete requirements to least complete requirements:

PM/ba or BA/pm as the PM
PM/BA and BA/pm as co-PMs
PM/BA and BA/PM as co-PMs

Number of Departments Affected As soon as more than one department is involved in a project, a whole new set of dynamics arise that were not present when only one department was involved. These dynamics arise because each department has its own opinion and reason for doing one thing instead of another. That creates conflict between departments. There are multiple clients, and they don't all have the same opinion about what is needed and what should be done. The BA and PM must resolve those differences. Refer to Chapter 1 for a discussion on approaches to resolving the conflicts. These conflicts begin with requirements gathering and continue with choice of project management model, project planning, release scheduling, change management, acceptance testing, and installation of deliverables.

If those multiple departments operate independently of one another when it comes to processes, the problem is even more challenging for the BA and the PM. To what extent must the disparate processes be integrated into a single process, and how? Is it just business unit processes that must be integrated, or is it also project management processes?

Wal-Mart provides an interesting example. Its 285 client groups are each supported by a project management process that was designed to exactly fit the needs, practices, and processes of the client group. There was no reason to believe that one client group's project management process was at all comparable to the next client group. I know from actual experience that they weren't. All is well until two or more client groups had to work on the same project.

The Wal-Mart solution was to form a "Core Team" to work on the project. The Core Team members are the recognized SMEs from all the affected business units. Their role is to:

- Advise each business unit's team members on technical matters.
- Provide SME help on enterprise systems and processes.
- Support each business unit team as requested and as needed.
- Collaborate with and advise the PM as requested.
- Negotiate and help resolve problems.

You can get more details on this unique approach by reading about Core Teams in my book *Effective Project Management: Traditional, Agile, Extreme,* 5th ed. (Hoboken, NJ: John Wiley & Sons, 2009).

The problems become heightened if the project involves departments from different organizations.

STAFF ENVIRONMENT Team size and availability are the two major factors that describe the staff environment. Both have an impact on the management staffing of a project.

Team Size As projects require larger teams, the team organizational structure expands to include more intermediate management levels. These will either support the project management process or manage the product/process of the project. In either case the co-PMs will both have to occupy positions at the PM/BA and BA/PM Senior Manager levels.

As PMs on a project, the PM/BA and BA/PM responsibilities are to manage the work, not the workers.

For the largest of projects, the management structure will change. The co-PM role becomes more of a coordinating and facilitating role. That manager can be either a PM/BA or BA/PM Senior Manager. One will be a better fit for the project than the other, and the choice will be clear from the nature of the project. The deliverables will be decomposed into several subprojects with each one managed by a PM/ba or BA/pm Manager as appropriate. That decomposition can be continued to sub-subprojects and the appropriate adjustments made to the management staffing of those sub-subprojects.

Availability Availability of appropriately qualified staff is always a problem. The best project management level personnel are in greatest demand. If projects can be prioritized according to complexity and PM/bas and BA/pms can be prioritized according to effectiveness, then you will have a way to appropriately assign managers to projects without wasting resources or assigning someone to manage a project for which he or she is not fully qualified or is too qualified. I think it would be better to postpone a project rather than to assign a less experienced PM just because the person was available.

REMEMBER:

Associate Manager Level: Responsible for small projects or major activities within a larger project
Manager Level: Responsible for large projects
Senior Manager Level: Responsible for very large or complex projects or programs

ORGANIZATION ENVIRONMENT The issue here is more one of organizational maturity and support of the project environment. As projects become more complex and uncertain, the organization's project infrastructure needs to become more comprehensive in order to support the projects effectively. If that support is not there, these challenging projects will be even more challenged, and failure rates will skyrocket.

Project success does not happen in the absence of investment in PMs and BAs, the infrastructure to support them, and executive management support and endorsement.

Project Management and Business Analysis Office Chapter 7 discusses the Business Project, Program, Process and Portfolio Support Office (BP^4SO) that I recommend for the development and support of the PM and BA professionals. The BP^4SO is the clearinghouse through which PM and BA management assignments are made and consultants assigned to project teams on an as requested basis.

The BP^4SO is the best linkage between projects and the management staff who are assigned. A good project portfolio management system is essential, especially one that is based on human resource constraints to deciding what projects will be staffed and how they will be staffed.

Maturity of the Business Analysis and Project Management Processes and Practices If processes and the practice of those processes are mature, then management assignments can be made with confidence. If that is not the case, then some additional uncertainty and risk is introduced into the project. That risk has to be mitigated, and one way to do it is through staffing the management of the project with higher-level positions than otherwise might have been done.

The organization has a big responsibility here, and that is to ensure that that maturity is in place and there are continuous process/practice improvement programs in place to maintain that maturity.

Human Resource Management System There need to be training, development, and career advising programs in place to keep the inventory of management-level PMs and BAs aligned with the demand for those staffs. That means also having a process in place to forecast the number of projects of each type that will be proposed for inclusion in the portfolio.

Project Portfolio Management System The strategic plan of the organization is input to the project portfolio management system. The strategic plan is the guide to the kinds of projects that should be supported to maximize the expected business value relevant to the strategic plan that the portfolio will produce.

MARKET ENVIRONMENT To be successful and contribute lasting business value, a project must be responsive to changes in the market. The economy in general can impact the project, but actions of competitors tend to be more of a factor. In general, if the market is a major controlling factor, then the project needs a BA/pm or BA/PM focusing on its process and/or product. Management responsibility for the project belongs with a PM/ba or PM/BA. The PM's knowledge of business analysis will be helpful because he or she may need to turn on a dime to keep the project aligned with delivering acceptable business value. A co-PM structure is a good investment because it supports the collaborative environment that will be important to the success of the project.

Opportunities from the PM and BA Integration

The major opportunity that this integration presents is to favorably impact project success. We know that the top reason for project failure is lack of client involvement. Permit me to change that reason to lack of *meaningful* client involvement. We've had client involvement for years, but it was never meaningful involvement. It was more like token involvement. The relationship among the client, the project, and the project management team is critical to supporting that involvement. Management staffing decisions must be decisions that are most supportive of engendering meaningful involvement.

Mapping the PMs and BAs into the Project Landscape

Based on what we have discussed so far, those position types that are simultaneously qualified to manage a project or manage business analysis activities or both simultaneously map into the project landscape as shown in Figure 6.2.

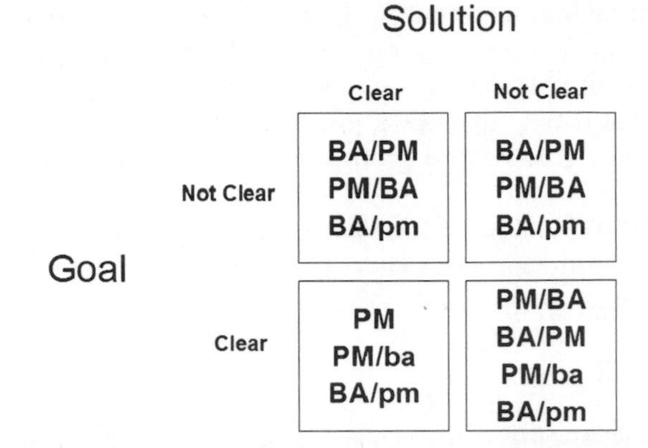

FIGURE 6.2 Project Manager Position Types Mapped into the Project Landscape

Within each quadrant, there is more than one position type that could provide candidates for managing the project. The actual choice of which position type is a best fit for managing the projects in that quadrant depends on the four factors described in Table 6.1. Let's take a closer look at the projects in each quadrant and how those factors will influence the choice of project manager.

TPM Project Managers (Q1)

Some of the TPM projects will be very simple. Perhaps they are projects that have been done repeatedly and for which a template work breakdown structure (WBS) exists and can easily be modified for the project at hand. There might be a parts list that can be adjusted accordingly. A well-defined risk management plan has been honed over the years with previous similar projects. The skills that the project team will require are established, so staffing is straightforward. Many of the team members will have worked together on previous projects so their work styles and habits are known by the team. For example, installing a network in a field office would fit these conditions. Everything is defined and ready to start. These projects can be managed by a PM or PM/ba at the Associate Manager Level. Conversely, a BA/pm will have the minimal project management skills to manage these simple projects effectively. A BA/pm with a career goal in the BA/PM or PM/BA sectors would be well served by managing such projects. The business analysis content of these TPM projects is minimal. The assignment draws more on the BA/pm's project management skills than business analysis skills.

Projects that will be managed using a linear TPM model will be candidates for having a BA/pm manage them. Projects that will be managed using an incremental TPM model are best managed by a PM or PM/ba. A BA or BA/pm assigned to such projects should be focused only on the business analysis activities of the project. Leave the management of the project to the PM or PM/ba.

APM Project Managers (Q2)

APM projects can range from those whose solution is almost complete and alternatives to complete the solution are available to those projects whose solutions are almost totally unknown. Projects at this level of complexity may have been attempted before without a satisfactory result. So APM projects have a significant range of complexity and uncertainty. To be effective in these projects, the PM must have a strong resume of successful APM experiences.

The product/process content of the project will determine whether a PM/ba or PM/BA is a better fit for managing the project. APM projects can benefit from having co-PMs; for example, one manager could be a PM/ba and the other could be a BA/pm. For very complex APM projects with high uncertainty, where little of the solution is known at the outset, the co-PMs should be a PM/BA and a BA/PM at the Senior Manager Level.

xPM and MPx Project Managers (Q3 and Q4)

All of the projects in these two quadrants are high-risk projects and require xPM projects. Much of the risk can be aligned with "looking for solutions in all the right places" or undertaking a project that has no solution that is acceptable from a business value standpoint. The best mitigation strategy for these situations is to recruit as creative a project team as available and stay out of their way. The project team is a major determinant of project success, and there should be no barriers to team members exercising their creativity. Depending on the criticality and importance of the project, the PM should be at the Manager or Senior Manager Level. Projects whose work plan is fairly well defined could be managed with a BA/pm. At the other extreme, when the work plan is not clear and the project content is quite uncertain, a BA/PM or PM/BA at the Senior Manager Level would be a better choice to manage the project. These are the majority of the projects in the xPM or Emertxe Project Management (MPx) quadrants.

The remaining question is whether the PM/BA or BA/PM who is managing the project will also be responsible for managing the BA activities on it. Someone on the team must be an SME for the product or process under consideration. That could be the BA/PM or PM/BA professional who is also responsible for managing the project. The conditions under which one person should have responsibility for both roles are:

- The BA content is the major activity in the project. such as some APM and most xPM projects.
- Scarce PM resources.
- Small and very simple TPM projects.

I think the major consideration is the skill profile of the available manager candidates.

When Should One Professional Have Both Project Management and Business Analysis Responsibility on the Same Project?

I can now offer my answers to the question that has been uppermost in many of your minds and has been the subject of many blogs and comments in published articles. There are several opinions, and I'll try to reflect them here. You should have already guessed that the best answer is "It all depends." I have established the need for professionals to be skilled in both disciplines, so we know that we already have professionals who are properly skilled to assume these dual roles on a project if the situation dictates. Let's identify the conditions under which it makes sense to make such assignments.

I have already discussed the co-PM model. In a sense, both the PM and the BA already have that dual management role, but it is shared equally. They are both responsible for the project outcome. Except for simple TPM projects, I have recommended that a dual role model be used. I have used that model for every project for over 20 years in my consulting practice, and it has contributed significantly to ownership of the project by the client and the BA, meaningful client involvement, and improved project success. Any obstacles to solution implementation have been neutralized just because meaningful client involvement has resulted in clients owning the solution. They were full partners in creating the solution so they are fully vested in its successful implementation. In a sense, their reputation and credibility are on the line.

Let me be clear that I am talking about management responsibilities. In the co-PM model, both the PM and the BA are responsible for jointly managing the work of the project with a focus on their discipline. They do not *do* the work. They *manage* the work. The PM has always been responsible for managing the work. If the project is large enough to justify the use of a BA in this co-PM management role, that person manages the business analysis work. Other BAs at the Associate Manager or Team Leader levels will do the work. For the BA in a management role, this may be new.

> *In the co-PM model, both the PM and the BA are responsible for jointly managing the work of the project with a focus on their discipline. They do not* do *the work. They* manage *the work.*

In the model where there is only one professional responsible for managing the project and the business analysis components, he or she again manages the work, as in the co-PM model. Only the smallest and simplest TPM projects typically will use this model. In larger or more complex TPM projects, the project management component and the business analysis component are managed separately by a PM and a BA in a co-PM model. These co-PMs should come from the Manager Level in the PM/ba or BA/pm sectors as a minimum. For some TPM projects, this will be sufficient; for larger TPM projects, the co-PMs should come from the Manager Level in the PM/BA and BA/PM sectors.

APM projects present a very different set of circumstances. They will all use a co-PM model. The more complex and uncertain the project, the more these co-PMs should come from the PM/BA and BA/PM sectors at the Manager or Senior Manager levels.

For xPM and MPx projects, the co-PM model should be used, but the focus is on product or process rather than following a scripted project management approach. My recommendation is that the co-PMs should come from the Manager or Senior Manager levels in the BA/pm, BA/PM sectors first and the PM/BA sector as a last resort. The project management process should not compromise the creative environment needed for project success. If the PM has more of a BA orientation, that will help.

Staff Availability

All of this is well and good, but in the final analysis, availability may control the decision about who manages the project. Remember that availability is not a skill and we don't assign a BA/PM or PM/BA Senior Manager to be responsible simultaneously for the roles of PM and BA on the same project just because such a manager is available and there are no other management choices. I would rather delay a project than make a less-than-acceptable management choice and assume more risk on a project that is already risky. As you move from APM to xPM to MPx projects, risk goes up just because of the type of project. The best management choice will be the one that adversely affects project risk the least. But I wouldn't compromise project success just to get a manager in place. The same holds true when a co-PM model is the best choice but the appropriately skilled and experienced PMs and BAs are not available. As the risk due to project type increases, the choice of delay versus inappropriate management staffing favors delay. The pressure to get an APM project going can be great because of the criticality of finding an acceptable solution quickly. That pressure to start often results in staffing a project with staff who are available rather than staff who have the requisite skills.

Team Size and Structure

As project size increases, team size increases and the management structure of the team deepens and broadens. The work gets pushed farther down in the team structure, and the management responsibility for the work gathers at the top. Several layers of management both from the PM and BA sides of the project are put in place. While a co-PM model can be used, senior management usually opts for a single PM to have ultimate responsibility for the project. One-stop shopping makes sense for reporting up into the organization. As far as managing down into the project, I would put my money on the co-PM model. For very large and complex projects, the single Senior Manager PM/BA or BA/PM at the top might vest responsibility in a PM and a BA who equally share project responsibility within their own discipline. I have been successful using this arrangement. I have found that it promotes better decision making when there are two constituencies voicing and defending their opinions, especially when the opinions differ.

Client History

If there have been problems with meaningful involvement with this client in projects in the past, I recommend a project management structure that is closer to the client. For example, a BA/PM managing the project would give the impression of being closer to the client than a PM/BA, even though the two might have identical skill/proficiency profiles. With this management structure, consideration should be given to having the BA/PM responsible for both management of the project and management of the business analysis activities. Again, this will tend to minimize the client relationships with a one-stop shopping model for the client. I think this management model can mitigate the risks associated with relating to this type of client.

Generalist/Specialist

We know that the PM takes the lead in managing the process for defining and solving the problem and the BA takes the lead in solving the problem and installing the deliverables. If you hold to this, then it is obvious that both the PM and the BA are involved in the project from beginning to end. I made that point in Chapter 1. Refer to Table 1.2 and the discussion that followed. This brings up a discussion of the role of generalists and/or specialists on the project. Calling the PM a generalist and the BA the specialist is too simplistic. A detailed discussion of generalists versus specialists is beyond the scope of this book, but a few observations with respect to their roles on a project are in order.

ROLE OF THE GENERALIST ON THE PROJECT TEAM Both the PM and the BA have roles as generalists on the project team. In general (pardon the pun), the generalist's role is to view the project in the context of the organization. At the strategic level, generalists should make sure that the direction of the project is in alignment with the direction the organization has defined in its strategic plans. At the tactical level, their role includes:

- Ensuring that the problem has been framed correctly and that its scope is not unnecessarily too limiting
- Viewing the problem and its solution in the context of other problems
- Looking for any learning from prior projects that they can leverage to the advantage of the current project

Every project must have at least one team member with this perspective.

PM Generalist By virtue of the fact that PMs will have broader experiences in the organization, they will be aware of related projects, client behaviors, management strategies, and solution approaches that may be of value to their current project assignment. BAs generally (there goes another pun) will not have this breadth of exposure and organizational experiences. PMs must ensure that in meeting the client's needs, the solution does not compromise enterprise-wide processes.

PMs also are responsible for choosing the management process that will be used to find that solution.

Further, PMs are likely to have a broader and deeper understanding of the existing business processes and the systems that support them than would BAs. PMs are in the best position to bring these experiences and understandings o the current project. As generalists, PMs must ensure an enterprise-wide view of the problem. The solution needs to support that enterprise-wide view. Many BAs will not have an enterprise-wide view; rather, the scope of their view is constrained by the particular business function(s) they are responsible for supporting. BAs are the SMEs or specialists and must represent their clients' needs.

Typical BAs would have a much narrower view focused on the business process they support and a few directly related processes.

BA Generalist At the Consultant Level and above, BAs should be able to adapt across every business unit and every process. The broader and deeper their cumulative experiences at the Professional Level, the more they will add value to their consulting engagements.

ROLE OF THE SPECIALIST ON THE PROJECT TEAM Both PMs and BAs have roles as specialists on the project team.

The PM Specialist PMs are the SME for project management on the project team. PM specialists can keep the organizational context and impact front and center with the business unit they propose solutions.

The BA Specialist As specialists, BAs bring the deepest understanding of a specific business process involved in the project and can be the most effective linkage between PMs and clients. If more than one business process is within the scope of the project, more than one BA specialist may be needed. As generalists, BAs do not have SME capabilities for any particular business process; rather, they have a toolkit of templates and processes that can be applied to any business function. There are roles for both types of BAs. The type of project will dictate whether a BA specialist or generalist or both are needed.

DOES EVERY TEAM NEED GENERALISTS AND SPECIALISTS? It all depends on the type of project.

TPM Projects A TPM project needs specialists for the product/process or under consideration. A generalist is not needed.

APM Projects An APM project needs generalists to keep the options open for solution discovery and specialists in the product/process under consideration. Having both a generalist and a specialist focus on the project team is critical for process/product design, development, and improvement projects.

xPM and MPx Projects An xPM or MPx project needs specialists in the product/process area under consideration. That means the PM needs to have more business analysis expertise than project management expertise. In order of project management preference, I would select the sectors from BA/PM to BA/pm to PM/BA. The more critical the project, the more I would depend on a co-PM model.

Putting It All Together

You should now appreciate the complexity of the PM and BA interactions and the options for their assignments. Every question regarding their assignment to specific roles in a project is answered the same way: "It all depends." Even that answer can change as the dynamics of the project play out. The world does not stand still because you made a staffing decision and are now managing your project. Change is constant; don't think otherwise as you make staffing decisions. You don't know what changes will arise or when, and when they do occur, you don't know what impact, if any, they will have on the team structure and the future of the project.

CHAPTER 7

PM/BA Career and Professional Development

In the previous six chapters I suggested several scenarios where a professional with some combination of PM and BA skills is needed. I searched the major players on the Internet for the PM/BA positions they list. The results were surprising and are reported in this chapter.

With all of the above taken into account, what are the next steps? Where do BA/PMs and PM/BAs come from? What training and education is in place to support their development? What about career paths and the training and education needed to support successful career growth? This chapter presents a comprehensive career and professional development planning model.

Examples of PM/BA Job Opportunities

In an attempt to try to understand the job market for these hybrid professionals, I did a search of Monster.com, Dice, and HotJobs. My search criteria were for positions with Project Manager/Business Analyst or some minor variation of that terminology in the title. The result was nearly 2,000 advertised openings. The actual data I collected on November 14, 2009 was:

Monster:	1390
Dice:	204
HotJobs:	322
Totals	**1916**

I'm sure there is some duplication, but the number of novel listings still is significant. Many of these postings were for senior-level positions,

positions requiring specific industry experience (i.e., finance, health care, or manufacturing) or application experience (i.e., SAP, PeopleSoft, or JD Edwards) or some combination of industry and applications experience. Position titles varied. PM/BA was the most frequently used, but Project Management Analyst and Business Process Manager were also common position titles. The titles themselves don't indicate whether the position is a PM/BA or BA/PM position but the descriptions did. I was surprised at the results, especially considering the dire economic and unemployment predicament we are experiencing.

In this chapter we will look at how the PM/BA landscape is used as the foundation of a career and professional development program for individuals and their managers as well as how the organization can support these efforts.

Several examples of recently advertised positions that fit into either the PM/BA or BA/PM sectors are presented next. Most also include specific industry experience. All of these examples place responsibility for both the project management and business analysis components into the same position. That does not imply that both project and business analysis management are concurrent assignments on the same project. It simply states that the requisite skills for project management and business analysis are vested in the same position.

PM/BA Position #1

Currently, Derivatives Clearing operations group provides clearing and settlement of U.S. Listed Equity Index Options and Futures products within securities services. The business is looking to expand into certain Credit Derivatives products, with a concentration on Credit Default and Interest Rate Swaps. The candidate will play an integral leadership role in the project management and business analysis function of this new technology and operations solution.

Responsibilities Will Include:

- Serve as a member of the location reporting to the Head of Projects and Management Support.
- Management of the entire project life cycle for the new derivatives initiative.
- Act as subject matter expert on the derivatives clearing process.
- Partner with Head of Derivatives Clearing Operations to define future vision for Operations, including processes and organizational structure.
- Involvement in third-party vendor analysis of technology platform.
- Provide status updates on the progress of the project to stakeholders.
- Develop accurate and detailed business requirements with the support of the appropriate resources.

- Develop strong relationships with selected technology solution provider.
- Develop comprehensive and robust test plan/cases and coordinate test execution for new technology solution.
- Initiate change management to adapt solutions and resources to a rapidly evolving business.
- Liaison with other senior members of the team, directly supporting the sales and product areas of the business from an operational perspective.
- Create and execute project work plans and revise as appropriate to meet changing needs and requirements.
- Review and edit requirements, specifications, business processes, and recommendations related to proposed solution.
- Ensure that issues are identified, tracked, reported on, and resolved in a timely manner.
- Manage day-to-day operational aspects of the project.
- Effectively apply our methodology and enforce project standards.
- Minimize our exposure and risk on project.

Position Requirements:

- BA required, advanced degree considered a plus.
- Minimum 8 to 10 years financial industry experience with an emphasis on derivatives products.
- Project management and business analysis skills are required.
- Understanding of project management methodologies including project plans, risk mitigation, issue tracking, and communication plans.
- Prior experience in a derivatives operations environment is required.
- Knowledge of credit and interest rate derivatives products is desired.
- Prior experience in vendor request for proposal (RFP) process a plus.
- Self-motivation and demonstration of initiative.
- Detail oriented.
- Ability to develop and document processes and procedures.
- Experience supporting new business/business growth and creating operational capacity.
- Proven ability to manage multiple priorities while respecting tight deadlines.
- Well-developed written and verbal communicative skills supporting an ability to represent the group at senior levels.
- Display a keen sense of ownership, responsibility, and accountability.
- Proactive, motivated, and willing to absorb new concepts.
- Excellent analytical skills and problem-solving skills.
- Basic MS Project and Visio skills.

PM/BA Position #2

The Projects Team assists senior management in making strategic decisions about how it invests capital in improving the Operations plant and manages projects that cross Operations' functional areas. Additionally, the Projects Team manages cross-divisional, cross-product, and firm-wide initiatives on behalf of ISG Operations. The role is multifaceted, offering the opportunity to take ownership for, and contribute to, the success of projects through their entire life cycle. The role will also entail scoping new projects that may be handed off to other areas for long-term ownership. The successful candidate will work closely with ISG Operations groups as well as other stakeholders that support the bank. The role requires the candidate to have an understanding of system architecture principles, project management fundamentals, and process analysis and design best practices. The successful candidate must also be able to analyze the commercial impact of business decisions. The candidate needs to understand basic financial analysis metrics and should be comfortable performing financial and other quantitative analyses. The need to switch between strategy and detail and the balancing of work on multiple projects will make this role complex, challenging, and rewarding.

Responsibilities Will Include:

Business Analysis

- Identify appropriate data sources (both qualitative and quantitative) for analysis when presented with business issues.
- Source and analyze data to provide insight into issues.
- Draft recommendations to address business needs based on analysis.
- Define solutions (process-based, reporting-based, technology-based) to business issues.
- Implement processes, tools, reporting solutions defined.

Project Management

- Facilitate cross-functional groups to achieve common goals.
- Define analysis- and implementation-focused project plans for self and small teams.
- Track and report status on cross-divisional projects being completed by individuals across Operations.
- Work with the stakeholders to prioritize work and drive milestones.
- Relationship Management.
- Develop an effective network of relationships across Operations.
- Influence and collaborate effectively with project resources not in the same reporting line.

Skills Required:

- Seven years' process improvement, business analysis, and project management experience.
- Clear understanding of project management and business analysis approaches.
- Strong quantitative and qualitative analytical skills and a creative and flexible approach to problem solving.
- Keen attention to detail.
- Strong communication and influencing skills.
- Strong sense of ownership and accountability for work.
- Capable of setting direction and motivating team members outside the direct reporting line.
- Fluency with all MS Office tools, including Access, Project, and Visio.
- Ability to deal with ambiguity and define approaches to bring unfocused issues to resolution.
- Exceptional written and verbal presentation skills.
- Bachelor's degree.

Skills Desired:

- Consulting experience for financial services industry clients.
- Experience working in or managing a Project Management Office (PMO) overseeing a large portfolio of project.
- Lean six sigma green belt or equivalent.

PM/BA Position #3

The company is offering exciting opportunities in our Human Resource Information System (HRIS) department for experienced HR systems project managers/business analysts. We seek dynamic, results-oriented individuals with proven problem-solving and analytic skills. The person in this role will be accountable for driving the successful delivery of systems and process solutions in our Professional Development (Human Resources) group. The candidate will have the opportunity to lead and contribute to both PeopleSoft HCM and internal systems projects. Through collaboration and partnership with the worldwide HR team, this individual will ensure that exceptional customer service is provided to our employees. The successful applicant will be responsible for project planning and administration, determining and documenting business requirements, creating systems specifications, developing and executing test plans and scripts, and streamlining and improving current processes. Key focus areas include: PeopleSoft HCM, North American Payroll, Global Payroll (UK), PeopleTools and PeopleSoft custom applications, reporting solutions, functional analysis for upgrades, PeopleSoft ELM (LMS), and recruiting solutions.

Qualifications

- Demonstrated project management skills with proven leadership experience in all phases of the implementation life cycle in HR systems projects.
- Experience in PeopleSoft reporting, Data Warehouse solutions, and Information Needs analysis.
- Experience in business analysis and execution of all steps in the system development life cycle in conjunction with implementation of PeopleSoft functionality.
- Expertise in Human Resources business processes.
- A minimum of seven years Human Resources systems implementation.
- A minimum of five years experience with PeopleSoft HCM 8.0 or above.
- Excellent written and verbal communication skills.
- Proven ability to thrive in a fast-paced team environment.
- Knowledge of PeopleSoft 9.1.

What Is a Professional Development Program?

In order to qualify for these senior-level positions, PMs and BAs will have to plan for their career and professional development. That requires a concerted and diligent effort with the support of one or more mentors. I'll call this plan a Professional Development Program (PDP).

A good PDP will answer these questions:

- Where am I?
- Where do I want to go?
- How will I get there?
- How am I doing?

The PDP planning horizon usually spans a full year with reviews on a quarterly basis or as needed. The temptation is to schedule them in conjunction with an annual performance review process. That is not a good idea. In practice, it is usually better to keep the PDP and performance review processes six months out of phase with one another. Having a performance evaluation and career planning session done concurrently isn't advisable. The performance review is not safe harbor for people. In a performance review, they are under pressure and in a defensive position. In a career planning session, they are in dream mode and don't need any additional pressure from the discussion of some other activity during sessions.

A good PDP will address these questions:

Where am I?
Where do I want to go?

How will I get there?
How am I doing?

WHERE AM I? This part of your PDP defines three cells:

1. **The highest level cell whose skill profile you exceed**
The most important part of this description is your skill/proficiency profile. These are not just with reference to the skills needed or used in carrying out the tasks and assignments of your current position. Rather, the skill profile covers all skills. Just because your skill profile exceeds the required skill profile for, say, the Senior Manager PM/ba cell doesn't mean your current position occupies that cell.
2. **The cell of your current position title**
You may not be in the best-fit cell. In the final analysis, it is a matter of supply versus demand and the nature of your assignment. For example, you might have the appropriate skill profile for a Senior Manager PM/BA position, but there are no openings for such a position. Instead, you are in a position in the Manager PM/BA cell. The more important question is: Where do you want to go? That is answered in another part of your PDP.
3. **The cell of your current assignment**
Continuing with the previous example, your position title is in the Manager PM/BA cell. Depending on a variety of circumstances, you could be assigned to a position in any one of these cells: Senior Manager PM/BA, Manager PM/ba, Manager PM/BA, Senior Manager BA/PM, Senior Manager BA/pm, Manager BA/PM or Manager BA/pm. The circumstances surrounding your current assignment could be such that you actually are occupying a position in a cell whose profile you greatly exceed or don't meet.

WHERE DO I WANT TO GO? While it may seem trite, I like to ask everyone who comes to me for career advice: "What do you want to be when you grow up?" The answers are amusing:

I don't want to grow up.
Employed.
I want your job.

In all seriousness, this is a basic question that must be answered before any career plan can be built. The answer isn't a lifetime decision either. Rather it is the place you would like to be professionally based on what you currently know and understand and what the business situation is likely to be as your career and job preferences mature. You can change your mind every Tuesday if you like. The important thing is that you have a career goal and a plan to work toward it.

The process for answering the question is through a relaxed conversation with one of your mentors. Your manager should be one of your mentors. Your manager is your best source for understanding where the opportunities lie for your short-term career prospects. And you should seek them out for that advice. Longer term prospects are best commented on by those who have a strategic perspective on the future and where the staffing opportunities lie. One of your mentors might be a person who occupies the position you seek as a career goal. You should seek their advice for that.

Both your manager and any senior-level manage whose advice you seek should be able to translate your career goals into a cell or sequence of cells that form a career path eventually leading to your ultimate career goal.

HOW WILL I GET THERE? At this point you know where you are (the cell you are in) and where you want to go (the cell of your career goal). Your next step is to build a career path for getting there. In fact, you could build several career paths that all lead to the same destination. One of those many career paths will be the one you pursue in greater detail.

And that is the answer to this question. The answer, however, will be given one step at a time. That step will be the PDP for the next planning cycle—probably 12 months. During that time you will put a four-part PDP together.

The PDP consists of:

- Experience acquisition
- On-the-job training
- Off-the-job training
- Professional activities

They are each described later in this chapter. Typically you will have reviews of the progress you are making. Those are often quarterly review sessions with your manager. The PDP should be very detailed for the next quarter and have lesser detail for the outer quarters.

HOW AM I DOING? Your PDP should be written so that it can be used to track progress. Just as you would develop an acceptance test in the form of a checklist, so also should your PDP performance tracking be in the form of a checklist. These checkpoints are specific accomplishments established in the PDP and progress towards them should be monitored on a quarterly schedule.

In addition to tracking progress against the plan on a quarterly basis, also consider revising your plan on a quarterly basis. Doing this could be as simple as taking advantage of an opportunity that wasn't known at the last checkpoint and that you would like to include in your PDP. Or it could be a little more comprehensive and focus on changing your career path or even your career goal.

PDP Contents

A PDP is the heart of every career and professional development effort. Your organization has a major role to play in PDP implementation and support; that topic is discussed in the next section. A PDP should contain a long-term career goal and a short-term career goal. The short-term career goal should cover the next planning horizon—most likely annual. It will be a detailed description of what is planned in order to achieve the short-term goal. The long-term career goal will be less detailed.

By way of introduction, I think that a good PDP is a plan developed by you and your coach/mentor/manager and consists of four components:

1. Experience acquisition
2. On-the-job training
3. Off-the-job training
4. Professional activities

Every position in every cell will have a minimum skill and competency profile required for it. These are the target profiles your PDP will aim for as you plan your skill development activities and the resulting career advancements. To qualify for a specific position, you must first define the skill and competency gap between your current and desired position and then build a PDP using the four components to remove that gap. That would position you to move to the desired position when a vacancy arises. Your mentor(s) should be available to help with plan development and other career advice.

Experience Acquisition

This part of the PDP describes the acquisition of further experience, mastering the skills and competencies needed in the current position. The further experience is also related to qualifying for entry into a higher-level position in the same or a neighboring sector. There may be certain areas that should be the focus of that additional experience—perhaps an area of noted weakness that needs improvement. So this is a do-more-of-the-same part of the PDP regarding the experiences noted above .

Here are examples of experience acquisition as they might be represented in the PDP:

- Seek out project assignments that have more of a business analysis focus than you have been doing.
- Support professionals who are more senior to you and have a business analysis skill that you need to improve to better meet current position requirements.

On-the-Job Training

First of all, the word "training" does not imply attending a formal course. Training can be very informal. For example, it may be attending an activity facilitated by an expert in the skill area of interest and observing what happens. Watching how someone does something is a learning experience.

On-the-job training is undertaken to increase the proficiency of skills and competencies needed to improve performance in the current position. If performance in one or more skill areas is below a nominal performance level, then training may be required to bring performance up to acceptable levels. Training need not be expensive and certainly does not have to be time consuming. There are several opportunities right under your nose. For example,

- Offer to help a colleague with a task to improve your skill to perform that task on your assignment.
- Attend a workshop to improve a current skill you are using on your job.
- Volunteer to join a project team that will further challenge you to take your skills to the next level.

You just have to pay attention and be on the outlook for opportunities. For example, John is a BA/pm Manager and has just been assigned PM for the Order Entry/Fulfillment Process Improvement Project. John has limited experience managing improvement projects, and he isn't too confident he can handle this project. Several past projects to improve the Order Entry/ Fulfillment Process met with little success. John's career goal is to become a BA/PM Senior Manager and eventually a Consultant to the Order Entry/ Fulfillment Process. So he asked Michelle, a PM/BA Senior Consultant, to help him improve his project management skills from the perspective of the BA on a process improvement project. Here are two examples of on-the-job training as they might be represented in the PDP:

1. Look for opportunities to observe and support the project management work of BA/PM professionals.
2. Take courses (on- or off-site) to enhance the project management skills required of your current position.

Off-the-Job Training

The purpose of off-the-job training is to increase skill proficiencies to the level needed to qualify for the next position in your career plan. In other words, this training is not relevant to meeting any current skill proficiencies associated with current job requirements.

For example, Harry is a BA Task Leader. His long-term career goal is to be a BA/PM Consultant. His short-term career plan is to become a BA/pm Associate Manager. Since his project management skills are very limited, he needs to get started on them. He is good friends with Larry, who is a BA/PM Senior Manager. Larry is respected among the BA community as one of the best project planners, and Harry feels that Larry has a lot to offer him. Larry is more than willing to be Harry's mentor and to help him learn project planning. Because he is so well respected for his project-planning skills, Larry is in high demand and always very busy. Harry approached Larry and asked to help Larry out in some way; in return, he could observe Larry. To Larry, that offer was a no-brainer, and he took Harry under his wing. Harry shadowed Larry and began learning project management from the ground up. When Larry retired, Harry took his place as a BA/PM Senior Manager and one of the organization's best planners. Harry is now one step away from his long-term career goal. Here are two examples of off-the-job training as they might be represented in the PDP:

1. Take courses (on- or off-site) to add business analysis skills that will be required by your targeted position in the PM/BA Associate Manager cell.
2. Look for opportunities to observe and support a professional practicing the business analysis skills you will need in your targeted position.

Professional Activities

This part of the PDP addresses participation in activities that are not necessarily related to any current or near-term career goal. Rather, they are just experiences that relate to your long-term professional interests, whether they are project management or business analysis related. They aren't necessarily directly related to these either. They might be related to better understanding your company's business. For example,

- Read the literature on your profession or your company's lines of business.
- Gain a better understanding of your competitors and how you might leverage your project management and/or business analysis expertise to improve your company's market share.

- Involve yourself in the Project Management Institute (PMI) and/or the International Institute of Business Analysis (IIBA) at the national and chapter level.
- Attend conferences and network with other PM and BA professionals.

Here are two examples of professional activities as they might be represented in the PDP:

1. Read books and journal articles on topics relevant to your targeted position in the PM/BA Associate Manager cell.
2. Attend meetings and conferences offering seminars and workshops relevant to your targeted position in the PM/BA Associate Manager cell.

A Deeper Look into the PM/BA Landscape

Each cell in the landscape will have a minimum skill and competency profile defined for all positions in that cell. In order for individuals to occupy a position in this cell, they must possess the minimum skill/proficiency profile for the cell as well as any additional requirements for the specific position they have targeted in that cell. For professional development planning, individuals will be in some particular cell based on their skill/proficiency profile and have career aspirations to move to another position in the same cell or to a position in another cell (usually this will be an adjacent cell). The skill/proficiency profile of the current and desired positions or cells can be compared, and the differences will identify the skill/proficiency gaps. The training and experience needed to remove that gap and be qualified to move to a position in the desired cell can be defined. The implications to the training department planning are obvious as are the applications to human resource management.

Short-Term Career Planning

The PM/BA Landscape is an intuitive tool that can be used to plan short-term and long-term career goals. Figures 7.1 through 7.3 illustrate using the landscape for short-term career goals. Figures 7.4 and 7.5 illustrate an example of a long-term career goal.

ASSOCIATE MANAGER PM/BA TO MANAGER PM/BA Figure 7.1 illustrates an example of a short-term career path that is very common and the simplest of the three examples of short-term career goals illustrated in Figures 7.1 through 7.3. The individual is seeking a promotion from Associate Manager to Manager in the PM/ba sector. The PDP for this short-term goal will focus on Experience Acquisition and Training that will contribute to current and future position. The type of project that the individual will

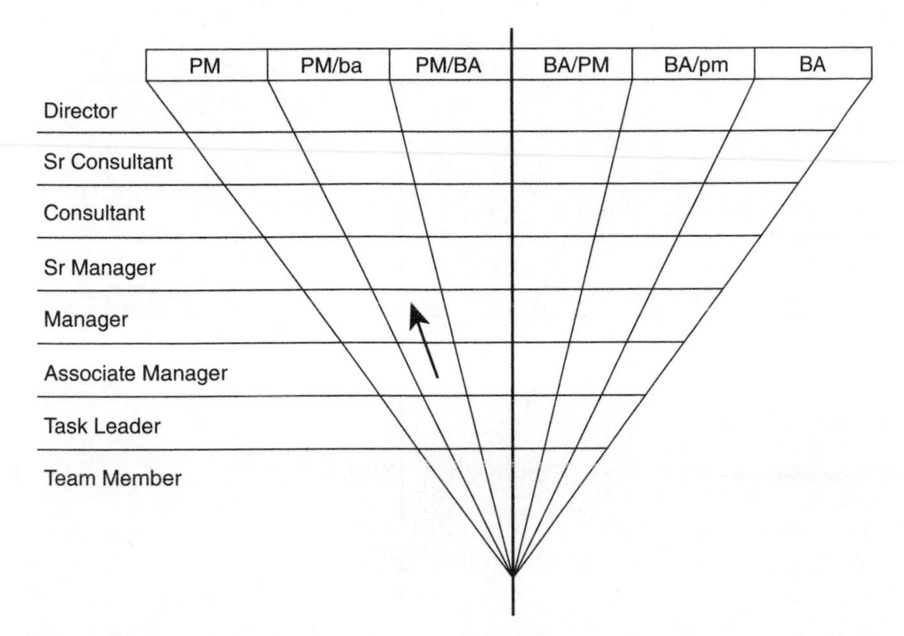

FIGURE 7.1 PM/BA Landscape Showing a Short-Term Goal to Seek a Promotion in the Same Sector

manage at the Manager Level is larger and more complex than those assigned to Associate Managers. In preparation for that change, a good experience strategy would be to get involved in more complex projects that are offered to Associate Managers.

ASSOCIATE MANAGER PM/ba TO ASSOCIATE MANAGER PM/BA Figure 7.2 illustrates a short-term career goal from a position in one sector to a position at the same level in another sector. Usually the change will be to an adjacent sector. While a short-term career goal to a nonadjacent sector is possible, it is not a good strategy. In my experience with this landscape, it is most effective when simple short-term goals are reflected in the PDP. A long-term career goal is a good idea and an example is given later in this section, but the short-term goals should be simple to plan and attain.

For the short-term career goal illustrated in Figure 7.2, the change is from the PM/ba sector to the PM/BA sector. The PDP for a sector change is more involved than the PDP for a promotion within the same sector. The beginning point for building the PDP is to analyze the proficiency skill gap between the individual's current business analysis profile and the minimum business analysis profile required for entry into the PM/BA sector at the Associate Manager Level. The business analysis skills can be prioritized and specific activities built into the PDP to remove those skill

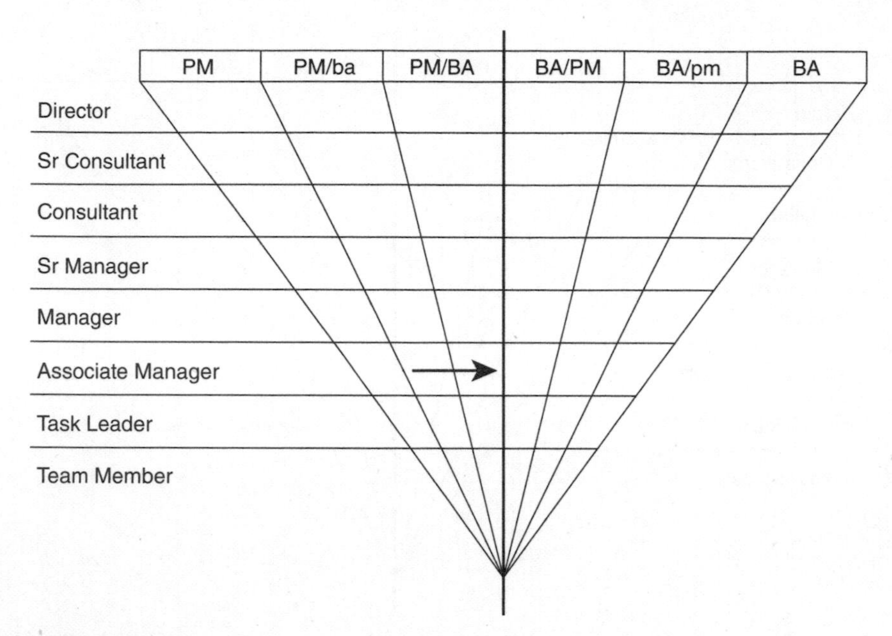

FIGURE 7.2 PM/BA Landscape Showing a Short-Term Goal at the Same Position Level in a Different Sector

gaps. Those activities will be a combination of both on-the-job and off-the-job training and professional activities.

This person is a professional project manager with basic business analysis skills and competencies. This is a very common position. Recognizing the importance and value of having stronger business analysis, the individual's short-term goal is to have a position in the PM/BA Associate Manager cell. To do this, the person will build a plan in the PDP to accomplish the short-term career goal. The PDP will focus on improving business analysis skill and competency profile from that of a PM/ba Associate Manager to that of a PM/BA Associate Manager. The PDP for this person might contain these strategies:

Experience Acquisition
Seek out project assignments that offer challenges to learn new business analysis skills or increase proficiency in existing business analysis skills. The best way to do this is to get more involved with the BAs assigned to projects you currently manage.

On-the-Job Training
Look for opportunities to take formal training in the skills for which your proficiency gap is the widest or for which there is a recog-

nized need in your project. Observing those who are well qualified in these areas is another way to improve your business analysis skill profile and contribute to the project teams you are on.

Off-the-Job Training

If there are business analysis skills for which you have no proficiency but which are needed to move from ba to BA competency for your next position, look for formal training to build your proficiency in those skills.

Professional Activity

A good reading program can broaden and deepen your knowledge and understanding of a business process important to your organization. That program might also include reading about a requirements-gathering approach that is popular in your organization for which you have no experience and little understanding.

ASSOCIATE MANAGER PM/ba TO MANAGER PM/BA Figure 7.3 is a more complex career path that incorporates the two previous examples into a single step. It simultaneously involves a position level increase and a change of sectors. The PDP for this short-term goal will have to include activities that contribute to the promotion from Associate Manager to Manager but

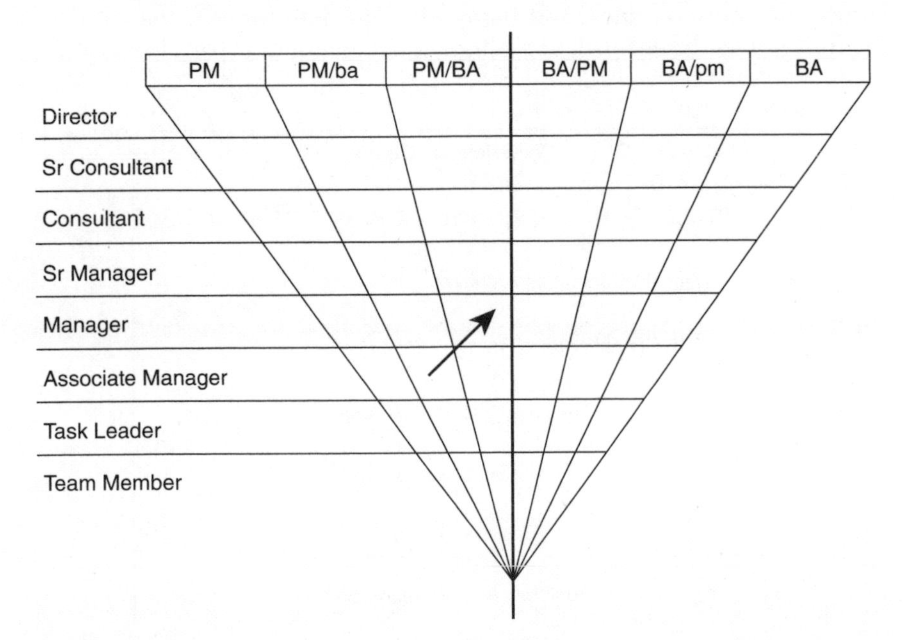

FIGURE 7.3 PM/BA Landscape Showing a Short-Term Goal to Seek a Promotion in a Different Sector

also the acquisition of the business analysis skills required of a Manager as compared to an Associate Manager. Achieving this goal may require more time than either of the simpler examples of Figures 7.1 and 7.2.

Rather than using this one-step strategy, a better option might be to follow one of these two-step strategies:

- PM/ba Associate Manager → PM/BA Associate Manager → PM/BA Manager
- PM/ba Associate Manager → PM/ba Manager → PM/BA Manager

It is likely that promotion opportunities may help you choose the better of the two strategies.

Long-Term Career Planning

The PDP should include both short-term and long-term career planning. As the short-term plan is executed, the long-term plan can be adjusted to take advantage of the situation. Even the targeted final position might change. Several factors will influence the plan and suggest revisions more compatible with the changing business environment and that offer more career growth and professional development opportunities along the way.

I've taken the data from Table 2.2, which gave an example of BA career path position titles, and mapped it into the career paths shown in Figure 7.4. One career path is the individual contributor path leading from

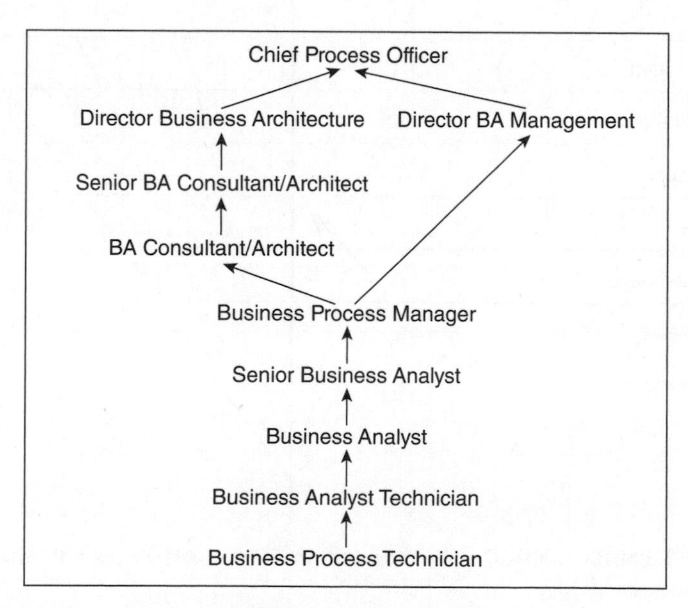

FIGURE 7.4 Generic Long-Term Career Paths

Business Process Manager to Consultant/Architect at the Professional Level through to Director of Business Architecture and finally to Chief Process Officer at the Executive Level. The other career path is a management path leading from Business Process Manager to Director of BA Management and finally to Chief Process Officer at the Executive Level.

This is still a generic career path within the BA profession. Business Process is a placeholder for a specific process, such as Supply Chain, Order Entry/Order Fulfillment, or Inventory Management. I'm not aware of this much definition and specificity in any organization, although it probably exists. The PM/BA and BA/PM position openings that I've seen advertised do not use business function specificity in their position titles, even though the detailed job requirements do focus on a specific function, such as finance or manufacturing.

The BA career path should follow a generic dual career path much like the one shown in Figure 7.4. I've applied that figure to the project landscape. The resulting mapping is shown in Figure 7.5. Note the dual career paths through the Professional Level and into the Executive Level.

CHANGE OF CAREER PATH Depending on prospects for future positions, there might be a better strategy for a career path. For example, if there is a defined shortage of BA/PMs at the Professional Level, the career path might be altered to improve the BA's promotion opportunities by adding

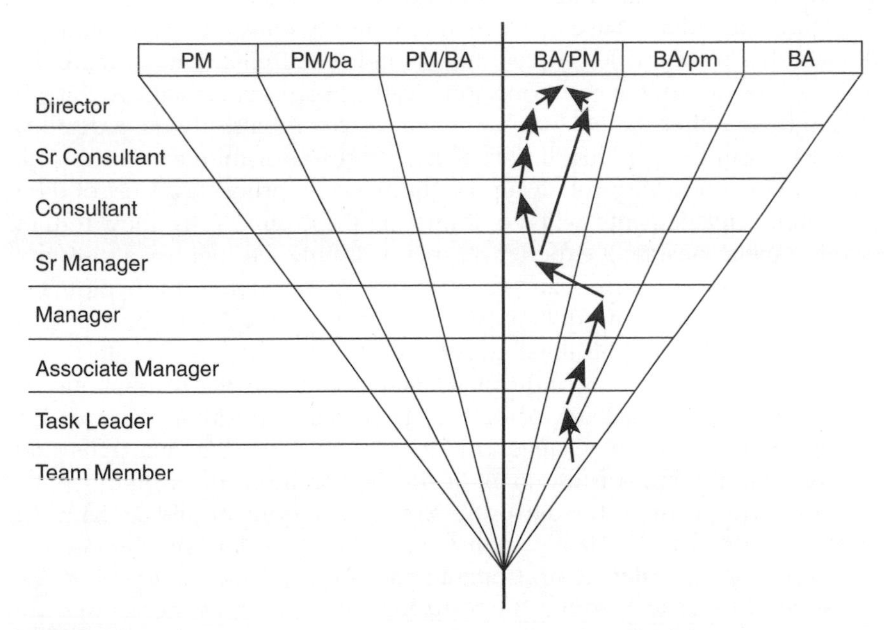

FIGURE 7.5 BA Career Path

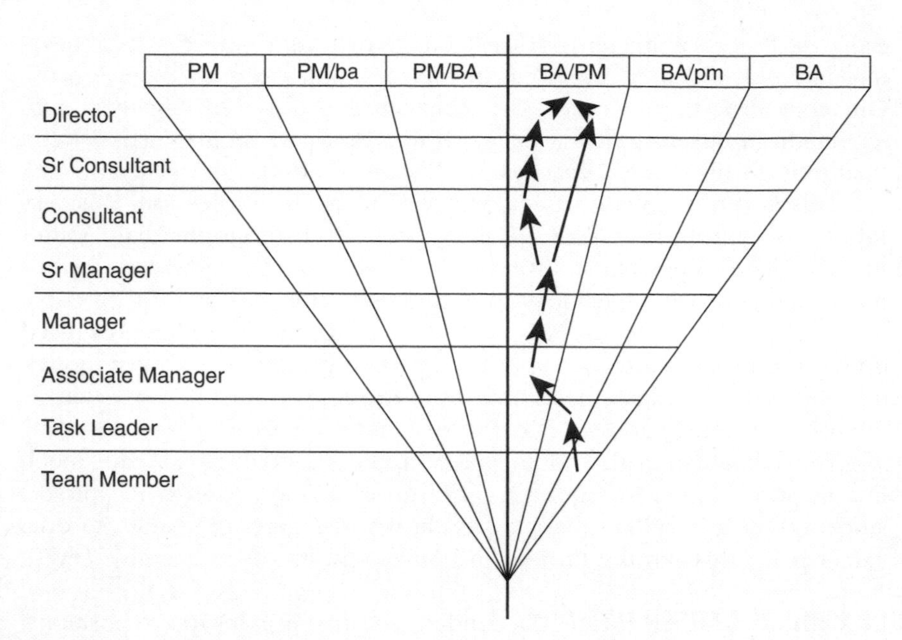

FIGURE 7.6 Modifying a Career Path to Take Advantage of PM Opportunities

project management skills faster than the career path shown in Figure 7.5. The revised career path might be something like Figure 7.6.

Once individuals have reached the Senior Manager Level position, a decision has to be made. To continue as individual contributors, the decision will be to follow the consulting career path. Consultant/Architects should be in either the PM/BA or BA/PM sector and already have detailed and demonstrated expertise in all of the processes within a business unit and working knowledge of all supporting business processes. Most of their consulting engagements will be found in these areas. As they further develop their consulting experience, they will broaden their expertise into other business processes and eventually be promoted to Senior BA Consultant/Architect. In addition, they can have a BA Architect role as a complement to their consulting expertise.

Rather than following individual career paths, they can continue as people managers in a Director of BA Management position. Director BA Management is my own suggestion for the position that manages and supports the BA human resources of the organization, much like a PMO Director is responsible for the professional development and assignment of PMs to projects in many organizations. The position Director of the Center of BA Excellence or Communities of BA Practice exists today and is the closest to Director BA Management, as long as these units are integrated into the formal organizational structure and are supported by executive management. As business process management matures, the Director of BA Management position will begin to appear.

In the final analysis, the best-fit career path depends on individuals, their ability to rise to a challenge, and the opportunities presented by their managers and the organization. The PDP can reflect this thinking but, in the final analysis, the PDP should only provide the detailed plan for movement to the next cell. Once a position in that cell is a reality, the detailed plan for the next cell can be formulated.

CHANGE OF CAREER GOAL A colleague of mine often says: "You will be smarter tomorrow than you are today" He is referring to improving task duration estimates, but his statement also can be applied to career planning. Today you might be a Business Process Technician and have ambitions to become Director of BA Management. That is a lofty and admirable goal, but a lot will happen before that goal can be attained. A Business Process Technician has much to learn about business analysis and project management to reach that goal. Along the way, several options will be presented, and several position opportunities will come and go. The goal may change several times based on the new learning that is acquired along the way.

FUTURE PROSPECTS No one can predict with any accuracy what the future holds for BAs and PMs as far as career opportunities are concerned. At this writing, the Executive Level positions in BA are almost nonexistent while there are a few such positions in PM (i.e., PMO Director and VP of Projects).

Many BAs believe that their career path leads from BA to Senior BA to a project management or product manager position. This demonstrates the misconception that exists in many organizations about the role of the BA. If BA is really destined to be a recognized profession, this misconception must be removed.

The first promotion in the BA family is to Business Analyst Technician. The incumbent will work under close supervision on the business analysis component of a subprocess or across a simple business process. The first Professional Level position is to a BA position, where the scope of work extends to a complete business process in a specific business unit like HR—for example, the Recruitment Process within the HR function. The next Professional Level position is to Senior BA, where the scope extends to all processes within a specific business unit, such as the processes that define the Human Resource Management System (HRMS) of the organization. From a Senior BA position, the common career progression is now to either project management or product management. I think that is unnecessarily limiting and offer instead the next level position as Business

Process Manager. In large organizations, this can be a significant position with responsibility for managing a hierarchy of lower-level Senior BA and BA positions. From the Business Process Manager position, professionals have to choose one of two directions for future career growth.

They can continue as individual contributors by following the BA Consultant/Architect and Senior BA Consultant/Architect career path. As part of their experiences as a Business Process Manager, they should have expanded their process knowledge and experiences into related processes, perhaps in other business units. This prepares them to pursue the BA Consultant/Architect career path to the Executive Level, such as Director of Business Architecture. This is a position that embraces Enterprise Analysis competencies at the highest level and has management responsibility for the entire internal consulting resources of the enterprise.

At the top of the BA career ladder is a new position: Chief Process Officer. As reported in the most recent BPTrends Report (Celia Wolf and Paul Harmon, *The State of Business Process Management—2010,* February 2010), this position exists in only a few organizations. A search of Monster, Dice, and HotJobs produced no hits on Chief Process Officer. The closest fit was a single hit on VP Business Process and Information Technology, but the position description didn't embrace business process leadership as I would have expected at this level.

TRENDS IN STAFFING NEEDS The project landscape is shifting away from the Traditional Project Management (TPM) quadrant toward the Agile Project Management (APM) quadrant. That puts the current supply of PMs and BAs out of alignment with the demand for them. In the short term, that limits the number of APM projects that the project portfolio can support. The temptation is to staff the growing number of complex and uncertain projects with PMs and BAs who are not qualified to manage such projects.

Organizational Support

Two areas need the support of senior management:

1. PDP
2. PMO, BA Center of Excellence, and BA Community of Practice

Professional Development Program

To be effective, the PDP needs the support of:

1. Your manager
2. Your organization

Your manager will probably be your mentor. But I have also seen examples where the individual chose a mentor who had risen to the individual's career goal position. A mentor could change over the life of your career, and you could have more than one mentor at a time. Keep that in mind too. You will look to your mentor to help you formulate your PDP and then manage it through those quarterly performance reviews. Since your manager is familiar with your PDP, he or she should make your work assignments supportive of your PDP.

Your organization can complement your manager in supporting the PDP. The four content areas of the PDP are a guide to the kind of support the organization should be offering. Here is a list by PDP part of what the organization could do to support the PDP effort:

EXPERIENCE ACQUISITION

- Keep a current list of open positions and make it available to all employees.
- Publicly post a list of activities of more senior staff in which the organization is willing to hire various support staff.

ON-THE-JOB AND OFF-THE-JOB TRAINING

- Publicly post a list of activities of more senior staff in which the organization is willing to hire various support staff or allow observers.
- Maintain an open archive of completed project notebooks for individual reference.
- Maintain a public listing of subject matter experts (SMEs) by discipline who are willing to provide individual advice, coaching, and mentoring.

PROFESSIONAL ACTIVITIES

- Support memberships in the appropriate professional societies in return for the individual's active involvement in the society.
- Maintain a corporate library.
- Offer to pay registration fees for conference attendance.
- Offer grants to support registration, travel, and expenses to professional conferences.

PMO, BA Center of Excellence, and BA Community of Practice

I am led to the conclusion that the organizational support of both the PM and the BA should lie in a single entity that I am going to call the Business

Project, Process, Program, and Portfolio Support Office (BP^{4}SO). Please excuse my taking liberties with the multiple use of "P." I do that for good reason. PMs have had the benefits of a PMO under various names for a number of years. The BA has not had a similar support organization. Recently we have seen BA Communities of Practice (BACOP) and BA Centers of Excellence (BACOE) emerging for the BA. Organizationally these arise in a number of ways. They can be very informal and provide forums for BAs to network and exchange ideas with peers, or they can be formal business units with goals, objectives, and a budget. Having formal organizational support units puts the PM and the BA on firmer footing in the organization's food chain. That is, they are visible, available, and promotable.

The organizational support of both the PM and the BA should lie in a single entity that I am going to call the Business Project, Process, Program, and Portfolio Support Office (BP^{4}SO).

In my opinion, there is no valid argument for not expanding an existing PMO to embrace both PMs and BAs. That is what I am calling the BP^{4}SO, and that is the focus of the next section.

Definition of the World-Class BP^{4}SO

The world-class BP^{4}SO is an enterprise-wide organizational unit that helps formulate and fully supports the tools, templates, and processes and the project managers and business analysts who use them to improve the effectiveness of the strategic, tactical, and operational initiatives of the enterprise.

The roles and responsibilities of the world-class BP^{4}SO are listed next.

- The BP^{4}SO is directed by a VP, who has a voting seat at the strategy table.
- Fully participates in the formation and approval of the business plan at all levels.
- Establishes the processes for and monitors the performance of the project portfolio.
- Has authority and responsibility to set priorities and allocate resources to projects.
- Establishes standards for the tools, templates, and processes used by BAs and PMs.
- Monitors compliance in the tools, templates, and processes used by BAs and PMs.
- Establishes an integrated PM/BA position family with defined career paths.

- Provides a complete program (training and human resource management) for the professional development of all PM/BAs.
- Ensures that the skill and competency profile of the PM/BA staff is sufficient for the realization of the project portfolio.
- Allocates PM/BAs to the approved portfolio of projects.
- Offers a full range of support services to executives, sponsors, and project teams.
- Provide an approved professional development program for all PM/BAs.

Performance metrics include:

- The project management and business process methodologies are assessed at Capability Maturity Model Integration (CMMI) CMMI Maturity Level 5.
- On average, the practice maturity level is at least CMMI Maturity Level 3.
- The project failure rate is less than 10%.

As you can see. the VP BP^4SO is an integral part of the enterprise from the strategy formation level to tactical planning to execution at the operational level. It is that unit's responsibility to make the most effective use of the enterprise's human resources toward the realization of the business plan.

Mission of the World-Class BP^4SO

The mission of the world-class BP^4SO is to provide a comprehensive portfolio of strategic, tactical, and operational support services to all project teams, sponsors and executives in order to ensure the delivery of maximum business value from the project portfolio.

Objectives of the World-Class BP^4SO

- Define a project life cycle with stage gate approvals.
- Design, develop, deploy, and support a comprehensive portfolio of tools, templates, and processes to effectively support all projects.
- Design and deploy a project review process to support project teams, and monitor compliance to established standards and practices.
- Establish a project portfolio management process to maximize the business value of project investments.
- Establish a decision support system and dashboard to support executive management's project decisions and provide for the timely monitoring of the project portfolio status.
- Design, develop, deploy, and support a comprehensive Human Resources Management System (HRMS).
- Design, develop, deploy, and support a continuous process improvement program for the BP^4SO.

How Do You Know You Need a World-Class BP^{4}SO?

Several symptoms suggest you might need a BP^{4}SO. I've listed those next with a few brief comments.

INFORMATION TECHNOLOGY, PROJECT MANAGEMENT, AND BUSINESS ANALYSIS ARE DEPENDENT FUNCTIONS To become an agile business, the organization must realize that information technology, project management, and business analysis must work collaboratively in order to deliver business value. That realization is an essential characteristic of business agility. There are several models for the actual structure of the BP^{4}SO. They range from centralized to decentralized and to a hybrid that I call hub-and-spoke models.

PROJECT FAILURE RATES ARE TOO HIGH The data are all too familiar to us. Reports show 70% of projects or more fail regardless of how failure is defined. That is simply unacceptable. Many of the reasons for those high numbers are known to all of us. Those that relate to business analysis include requirements elicitation, user involvement, clear business objectives, minimized scope, and firm basic requirements. Those that relate to project management include choosing the best-fit Project Management Life Cycle (PMLC), selecting the appropriate team, planning and scheduling, performance monitoring and control, scope management, and risk management. All of these can be favorably impacted as long as the required organizational support is in place.

TRAINING IS NOT PRODUCING RESULTS I am not aware of any systematic study of the root causes of training ineffectiveness. Possible causes are inappropriate materials, inappropriate delivery, no follow-through on behavioral changes after training, or no testing of skills acquisition. Training needs to be taken seriously by those who attend. Attendees must be held accountable for applying what they have learned, and there must be ways to measure that application. I am amazed at how many training professionals and curriculum designers are not familiar with Kirkpatrick's model. It is a classic. The interested reader can consult Donald L. Kirkpatrick's *Evaluating Training Programs,* 2nd ed. (San Francisco: Berrett-Koehler, 1998). In my experience, project reviews that are held at various milestones in the life of the project are excellent points at which to validate the application of training. If clear evidence isn't shown that training has been applied, some corrective action is certainly called for.

HR PROJECT STAFF PLANNING ISN'T EFFECTIVE Organizations need to do a better job of defining the supply of project management and business analysis skills and the demand for those skills by project type. A well-designed program is needed to match the supply to the demand

and to make better project staffing assignments. The BP^{4}SO is the best place for these responsibilities to be carried out.

INABILITY TO LEVERAGE BEST PRACTICES The BP^{4}SO is the best place to collect and distribute best practices. Project status meetings and project reviews are the places to identify best practices. The BP^{4}SO, through some form of bulletin board service, is the best place to distribute that information. In the absence of that service, the collection and distribution of best practices isn't going to happen.

LACK OF CONTROL OVER THE PROJECT PORTFOLIO Many senior managers don't know the number of projects that are active. They haven't made any effort to find out or even be selective of those projects that should be active. That behavior has to change if there is any hope of managing the project work in the organization. The BP^{4}SO is the clear choice for stewardship of that portfolio. At the least, it can be the unit that assembles project performance data and distributes it to the decision makers for their review and action.

NO CONSISTENCY IN PROJECT REPORTING Without a centralized unit responsible for the reporting process, consistent and useful reporting isn't going to happen. Again, the BP^{4}SO is the clear choice to establish the reporting structure and assist in its use.

TOO MANY RESOURCE SCHEDULING CONFLICTS Most organizations operate with some form of matrix structure. Resources are assigned from their functional unit to projects at the discretion of the functional unit manager. In such situations, resource conflicts are unavoidable. The individuals who are assigned to projects are torn between doing work for their functional unit and doing work for the project to which they have been assigned. None of this is news to you. One solution to resource scheduling conflicts is to use the BP^{4}SO as the filter through which project staffing requests and staffing decisions are made. The major benefit of this approach is that it takes the project manager off the hot seat and puts the responsibility in the BP^{4}SO, where it can be more equitably discharged.

GAP BETWEEN PROCESS AND PRACTICE This is a major problem area for many organizations. They may have a well-documented process in place, but without any oversight and compliance function in place, they are at the mercy of the project manager to use or not use the process. The BP^{4}SO is the only unit that can close this gap. The BP^{4}SO puts the process in place with the help of those who will be held accountable for its use. The BP^{4}SO, through project performance reviews, can determine the extent of that gap and put remedial steps in place to close it.

In the list below, check all the boxes that describe your organization.

- ☐ Project failure rates are too high.
- ☐ The project management methodology is not widely adopted.
- ☐ Scope change requests are out of control through the project.
- ☐ One resource pool is staffing multiple projects.
- ☐ There is lack of project management expertise in needed areas.
- ☐ Several vendors and contractors are used across projects.
- ☐ There is a need to consolidate reports and metrics.
- ☐ Time to market is a critical success factor.
- ☐ Total project costs are too high.
- ☐ The resource pool is not aligned with staffing needs.
- ☐ Training is not impacting project performance.
- ☐ The HR project staffing plan is not effective.
- ☐ The PM substitutes for the BA too frequently.
- ☐ The BA substitutes for the PM too frequently.
- ☐ PMs are not skilled in BA processes.
- ☐ BAs are not skilled in PM processes.
- ☐ You have trouble leveraging best practices.
- ☐ You don't have control of the project portfolio.
- ☐ There is no consistency in project status reporting.
- ☐ There are too many resource scheduling conflicts.
- ☐ There is a noticeable gap between documented process maturity and actual practice maturity.

8 or more checked boxes: Establishing a BP4SO highly advised
12 or more checked boxes: Establishing a BP4SO a necessity

FIGURE 7.7 BP^{4}SO Readiness Assessment

How do you know if you need a BP^{4}SO? Figure 7.7 provides some criteria that might help you decide.

On the surface, the world-class BP^{4}SO won't seem much different from the traditional PMO. The world-class BP^{4}SO offers an expanded portfolio of support functions as compared to the typical PMO, but if you look under the hood, you will see that I am proposing that there is a significant difference. Organizations that see the handwriting on the wall know that project management, program management, portfolio management, and business process management are all converging on a single strategic role with enterprise-wide scope. There needs to be a seat at the strategy table to represent these entities to the organization. The director of the world-class BP^{4}SO that I envision should occupy that seat. That person helps define strategy and, through the BP^{4}SO infrastructure, provides for the enablement of that strategy.

Organizational Structure of the World-Class BP^{4}SO

The three choices for an organizational structure are centralized, decentralized, or a hybrid of the two.

CENTRALIZED STRUCTURE Except in large organizations, the centralized structure is the model used for the PMO in most organizations. The PMO generally has a single hierarchical structure and may have a staff of permanently assigned PMs or a staff of revolving PMs. The BP^4SO would be similarly organized.

Strengths of a Centralized BP^4SO:

- Generally well funded
- Greater flexibility and fit for staff deployment decisions
- Easier to establish a common set of tools, templates, and processes
- Be able to offer more career path opportunities
- Can develop a highly skilled BA and PM resource
- Strategically involved in the enterprise
- More cost-effective use of BA and PM resources
- More opportunity to develop specialized skills and competencies
- Easier to leverage best practices across disparate business units

Weaknesses of a Centralized BP^4SO:

- BA and PM loss of credibility with the business units
- Services and support do not match specific business unit needs
- Service levels not responsive
- Staff schedule conflicts

DECENTRALIZED STRUCTURE In the decentralized model, any business unit that can justify it can establish its own BP^4SO. These could be division level or functional department level. There could even be a BP^4SO that served enterprise-wide needs. Each would have its own vision, mission, objectives, and support services.

Strengths of a Decentralized BP^4SO:

- Greater sense of BA and PM loyalty to the sponsoring unit
- Business unit has greater leverage over its BP^4SO
- BAs and PMs gain greater expertise in applying their skills to the business unit they support
- Services better aligned with business unit needs
- Better client relationship management

Weaknesses of a Decentralized BP^4SO:

- Fewer career growth and professional development opportunities
- Shifts in enterprise priorities may result in less-than-optimal disposition of BA and PM resources
- Business unit loyalty can compromise support of enterprise or cross-functional projects
- Obstacle to sharing of best practices across business units

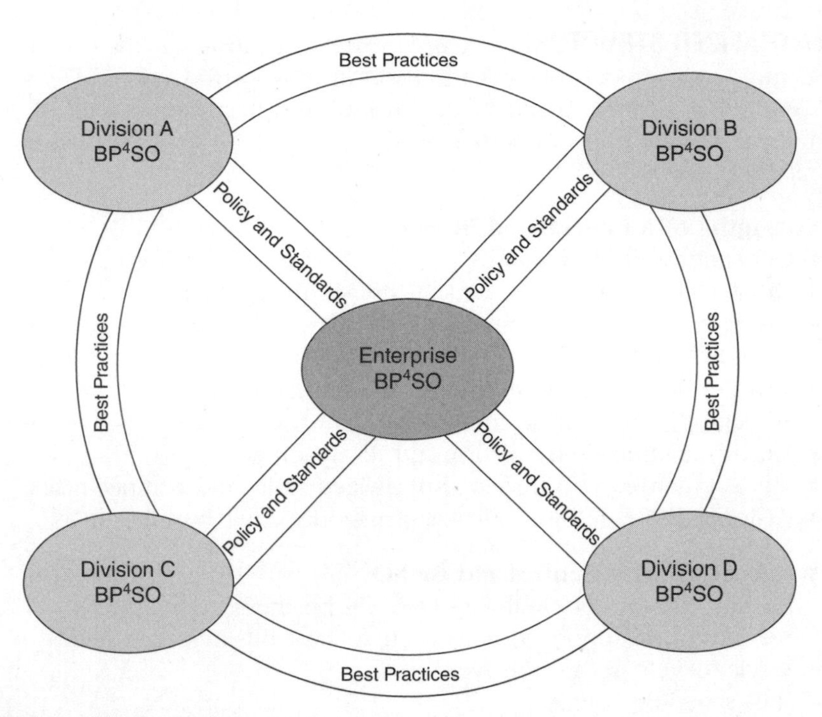

FIGURE 7.8 BP^{4}SO Hub-and-Spoke Structure

HYBRID STRUCTURE The only structure that makes sense to me is a hybrid structure, where the PM/BAs are close to their client groups but are loosely bound by enterprise-level policies and standards. That structure is what I call the hub and spoke and is illustrated in Figure 7.8. At the hub we have the enterprise-level unit that is responsible for policy, process, and staff development. At the spokes are the various divisions and departments with their own staff of PM/BAs. They might establish tools, templates, and processes specific to their needs but in compliance with the enterprise policies and processes.

The hub-and-spoke structure is fairly robust and can easily adjust to division differences. The Enterprise BP^{4}SO reports at the Executive Level and sets policies and standards for the enterprise. The Division BP^{4}SOs service its business units by adapting the enterprise policies and standards to the specific needs of the division. As long as the division adaptations are in compliance with the enterprise policies and standards, there are no issues. To assure that compliance, the development of the enterprise policies and standards should be done by a task force whose members are representative of the divisions. The Enterprise BP^{4}SO supports each Division

BP⁴SO, which in turn directly supports its business units. I like to think of the Enterprise BP⁴SO operates at the strategic-level unit while the Division BP⁴SOs are operational-level entities.

Strengths of a Hub-and-Spoke BP⁴SO:

- All of the strengths of the centralized and decentralized structures
- Sharing best practices is facilitated
- More cost-effective training programs

Weaknesses of a Hub-and-Spoke BP⁴SO:

- Requires good communications management
- Requires an added layer of compliance monitoring

Staffing the World-Class BP⁴SO

Five staffing models, discussed next, come to mind.

VIRTUAL BP⁴SO The Virtual BP⁴SO does not have any PM/BAs assigned to it. Instead they are deployed throughout the enterprise, where they are assigned on an as-needed basis by their organizational unit. These are not full-time project managers but are professionals in other disciplines who have project management skills and competencies as part of their toolkit.

The BP⁴SO staff consists of a manager and one or more assistants who support PM/BAs as required. They may be PM/BAs, but that is not necessary. The important thing is that this staff has the necessary expertise to provide the support needed. The support services may span the full list but are often quite restricted because of staff limitations.

PM/BAS ARE ASSIGNED TO THE BP⁴SO ON A ROTATING BASIS This is an excellent model for deploying the PM/BA methodology, skills, and competencies throughout the organization. In this model PM/BAs from the business units are temporarily assigned to the BP⁴SO as a sabbatical from being on the firing line too long. A sabbatical might last from three to six months, during which time they might conduct a specific process improvement project or simply act as mentor and coach to other PM/BAs.

PM/BAS ASSIGNED TO THE BP⁴SO In this structure the BP⁴SO PM/BAs manage critical mission or enterprise-wide projects. Usually not all PM/BAs report to the BP⁴SO. There will be several who are assigned to business units to work on less comprehensive projects.

PM/BAS ASSIGNED AT THE DIVISION LEVEL PM/BAs assigned at the division level work on division-wide projects. These may be projects that span two or more departments.

PM/BAS ASSIGNED AT THE DEPARTMENT LEVEL This assignment is the same as the division level except PM/BAs are assigned within a department and work only on projects contained within that department.

Implementing a World-Class BP^4SO

A world-class BP^4SO isn't born like Venus on the half-shell. Rather it is through a demand pull strategy. The first step is to implement the hub. That means choosing a Director and staffing the hub with a small number of PM/BA and BA/PM Consultant/Architects and a small support staff. Their initial responsibilities include defining policy and standards. They don't do that in a vacuum but rather by assembling a task force whose members represent the divisions that are able to justify a division BP^4SO. Once the policy and standards are in place, the division BP^4SO can be established by following a similar strategy. The focus at the division level will be the development of division standards that are compliant with the established enterprise standards. The division PM/BA and BA/PM Consultant/Architects will establish a task force from the business units in their division to define the division standards. As the demand for PM and BA support grows, existing division BP^4SOs can grow and new division BP^4SOs can be established.

Putting It All Together

In this chapter I have shown by example how the BA/PM landscape and BA/PM position family would be used in professional development and career planning.

The responses that I have received to previously published articles on this topic have been overwhelming, both positive and negative. Being a change management advocate, I am thankful for thought leaders' interest in PM and BA. I'm not surprised with your reactions. My hope is that we can continue the exchange. As always, I welcome opposing positions and the opportunity to engage in public discussions. Your substantive comments are valuable. Criticism is fine and is expected, but in the spirit of Agile Project Management (APM), so are suggestions for improvement. Also in the spirit of APM, I am trying to find a solution to the career and professional development of the BA, BA/pm, BA/PM, PM/BA, PM/ba, and PM.

I realize that I have taken a controversial position and in so doing have stepped out of my comfort zone and perhaps put myself in harm's way. I do so intentionally. Through the pages of this book, I hope I have gotten your attention. I want you to start thinking about the care and feeding of a single BA/PM and PM/BA professional—one who is fully skilled in both disciplines. I repeat again that these professionals do not automatically assume responsibility for both project management and business analysis at the same time on the same project. There are, of course, situations when that is the recommended approach, and I have discussed such situations. There will always be a collaborative relationship between the PM and the BA, and I contend that that collaboration will be more effective if they function as co-PMs and are PM/BA or BA/PM professionals. I have a very strong belief that there is a growing need for them. As you dig deeper into the dual-discipline professional position, I ask that you approach my suggestions with an open mind and offer your ideas directly to me at rkw@eiicorp.com or in the appropriate public forum.

In Conclusion

A Call to Action

This book has described a new professional—the PM/BA—and why we need such a professional, how the PM/BA position family and career path might be defined, and what a professional development program might look like. In essence, it is a blueprint for moving forward.

This closing comment focuses on the next steps that might be taken to bring the PM/BA professional into reality. My suggestion for the system that can provide for the career planning and professional development of PM/BA professionals follows. It is the logical next step.

A System to Prepare PM/BA Professionals

The project to design and develop this system will be a challenging project, but you already should have guessed that. Like every effective project manager, let's begin this project with what I call a Project Overview Statement (POS). For PMs, this is similar to the Project Manager Body of Knowledge (PMBOK) Project Charter. For BAs, this is similar to the Business Analysis Body of Knowledge (BABOK) Project Scope Statement. This POS will be the framework and guide for all project work to follow. For reference purposes, this project will be called the Professional Development Program (PDP) Systems Design and Development Project. This chapter does not discuss deployment. Deployment of the PDP system is left for another project and is briefly mentioned later.

Project Overview Statement

The POS is the first document that describes a proposed project. It is a high-level description of a business situation and what you propose to do about it. It is a one-page document with five parts. I have used it for more than 40 years with great success. For the PDP Systems Design and Development Project, here is my version of the five parts of the POS.

PROBLEM/OPPORTUNITY The project landscape is changing. Complexity and uncertainty dominate. Only rarely can requirements be completely defined and documented at the outset. An agile approach to these projects is highly recommended. To be effective in managing these projects, the agile project manager must be fully skilled in both project management and business analysis.

GOAL Design and develop an Internet-based career planning and professional development system to prepare professionals to be both project managers and business analysts so that they are fully capable of successfully managing projects at all levels of complexity and uncertainty.

OBJECTIVES

- Define the PM/BA position family.
- Define the PM/BA career paths.
- Identify the skills and competencies required of the PM/BA professional.
- Establish the minimum skill/competency proficiency profile of each PM/BA professional.
- Define the internet-based PDP.
- Design the skill and competency assessment tools portfolio.
- Design the career planning module.
- Design the professional development module.
- Design the integrated PDP System.
- Develop the integrated PDP System.
- Document the integrated PDP System.

SUCCESS CRITERIA

- The PDP System will be a thin client Internet-based system.
- The PDP System will be ready for deployment within 12 months of starting the project.
- The PDP System will be parameter-driven and fully support user-definable PM and BA position families, career paths, and skill/competency profiles.
- Using the PDP System will not require any training—it will be intuitive.
- The PDP System will not have a user guide.

ASSUMPTIONS, RISKS, OBSTACLES

- The need for a PM/BA professional will not be acceptable to the entire BA and PM communities.
- The project will be co-managed by a representative from the PM community and a representative from the BA community.
- Qualified technical resources will be made available when needed to design and build the PDP System.
- The BA and PM communities will be honest participants in reviewing and commenting on the PDP System .
- An Agile Project Management approach will successfully deliver the PDP System.
- A sponsor can be found to financially support the project.

Suggested High-Level Work Breakdown Structure

Once the POS has been approved by representatives from both the BA and PM communities, detailed project planning can begin. A high-level work breakdown structure might look something like this:

Phase I: Scoping the PDP System

1. Describe the PDP System Task Force purpose and membership.
2. Recruit the PDP System Task Force members.
3. Plan and hold the PDP System Kick-off Meeting.
4. Synthesize current BABOK and PMBOK position definition and documentation.
5. Document the requirements of the desired PDP System.

Phase II: PDP System High-Level Design

1. Define and document the PDP System deliverables.
2. Define and document the PDP System process flow.
3. Gain approval of the PDP System high-level design.

Phase III: PDP System Detailed Design and Documentation

1. Recruit the PDP System Development Team.
2. Design the documentation format and templates.
3. Construct the PDP System documentation.
4. Circulate PDP System documentation for review.
5. Revise PDP System documentation.
6. Gain approval of the PDP System detailed design.

Phase IV: PDP System Development

1. Recruit PDP System Development co-project managers.
2. Recruit PDP System Development Team.

3. Review the PDP System documentation.
4. Define the PDP System Technical Requirements and Architecture.
5. Prioritize PDP System Requirements.
6. Define PDP System Development cycles (plan, build, check) and time boxes.
7. Execute PDP System Development Cycles.
8. Demonstrate having met the requirements of the PDP System Task Force.
9. Discharge the PDP System Development Team.

Phase V: PDP System Marketing

1. Create the PDP System Marketing Program.
2. Plan and publish PDP System articles.
3. Design and produce PDP System promotional materials.
4. Distribute PDP System promotional materials.
5. Discharge PDP System Task Force.

A Call to Action

So there you have it! The complexity and challenge of the PDP System Design and Development Project should not be underestimated. Its importance cannot be overstated either. It is my firm belief that having PM/BA professionals on your staff and coordinated through a BP^4SO will have a significant impact on project success.

Testimonial data that I have gathered over the years from over 10,000 project managers worldwide suggests that over 70% of all projects are in the agile category. These projects are such that requirements identification and solution definition can come about only from learning and discovery during project execution. That requires the use of some form of iterative approach. This is clearly the domain of the agile project and requires the leadership of the PM/BA professional. That they are needed is not debatable. The processes to develop them are by no means obvious or in place.

Through this book I've tried to build the case for formally recognizing the need for the PM/BA professional and for the systems to meet their career planning and professional development needs. I've offered a high-level work breakdown structure as the beginnings of the project plan to put the requisite PDP System in place to support those needs.

How might we make that plan happen? If your organization sees the importance and the need for such a system and suffers the pains of frequent distressed projects and excessive project failure, perhaps it would be interested in funding a PDP Systems Design, Development, and Deployment Project to meet its own needs. This would get you off to a fast-track start. Could your employer be one of those companies?

Alternatively, a BA or PM product/service provider might be interested in adding the PDP System to its portfolio through a joint venture to design, develop, market, and sell the PDP System. Do you work for such a company?

Finally, the higher education and training communities should be paying attention and begin planning the curriculum and courses to meet the need for the merged project management and business analysis professional. Their lead time is substantial and time is wasting.

The next step for *me* is to partner with a BA/PM professional who would be interested in taking an advocacy position with me and then begin designing and developing the PDP System. If you and I are of like mind, if you are an accomplished BA professional, if you have name recognition in the BA community, if you have some time and would like to make a difference, I would like to hear from you and how we can collaborate. My direct email is rkw@eiicorp.com. I'm serious!

APPENDIX A

Skill/Proficiency-Level Matrices for the Eight PM/BA Position Levels in the Project Management Landscape

What we need is a matrix for each sector in the position landscape. That will require six skill/proficiency matrices—one for each sector. Each matrix has 8 columns—one for each position type. Each matrix will have 89 rows of proficiency data—one for each skill. The list of skills is identical for all six matrices. Each cell will contain the minimum proficiency level for the skill and position type that defines that cell. Every position that falls in a particular cell defined by position level and position type (shown by the cells in Figure 3.1) must meet the minimum proficiency levels established for that cell. Even though this data is quite detailed, I still consider it a guide and a baseline for an organization's consideration. For a specific position in a cell, these minimums can be adjusted upward.

These matrices did not exist when I started this book project. To build these matrices, I consulted the four major IIBA and PMI documents:

- International Institute of Business Analysis. *A Guide to the IIBA Business Analysis Body of Knowledge (BABOK Guide), Version 2.0.* Marietta, GA: Author, 2009. (ISBN 978-0-9811292-1-1)
- International Institute of Business Analysis, *A Guide to the IIBA Business Analysis Competency Model, Version 1.0.* Marietta, GA: Author, 2010. (ISBN: 978-0-9811292-1-1

- Project Management Institute, *Project Manager Competency Development Framework,* 2nd ed. Newtown Square, PA: Author, 2007. (ISBN: 1-933890-34-7)
- Project Management Institute, *A Guide to the Project Management Body of Knowledge (PMBOK Guide),* 4th ed. Newtown Square, PA: Author, 2008. (ISBN: 978-1-933890-51-7)

Neither the IIBA *BABOK* nor the PMI *PMBOK* offers any skill or proficiency data. These documents are standards guides, and skill profiling is not within their scope. Based on the PM and BA landscape defined in Figure 3.1 and my 40+ years of PM and BA experience, I have drafted the first definition of skills and proficiencies as reported in these six matrices. A few thought leaders in project management and business analysis have commented on the skills they recommend, and I have tried to be consistent with that research tradition.

Using Figure 3.1 as the structure, Tables A.1 through A.6 present the skill and proficiency levels for each combination of position level and position type. These are the minimum skill/proficiency profiles that make sense to me. You be the judge. In application, consider them as a starting point to be modified to fit a specific application.

I have used a six-category classification of skills in my own consulting practice for over 10 years. It has served me well and met with client acceptance too. I'm using that same classification scheme here.

1. **Project Management.** The skills essential to the definition, planning, execution, monitoring, modification, and closing of a project.
2. **Business Analysis.** The skills essential to six Knowledge Areas defined by the BABOK.
3. **General Management.** The skills that relate to the management of the enterprise.
4. **Business Functions.** The skills related to specific activities of the enterprise.
5. **Interpersonal.** The skills whose practice requires more than one person and are not linked with any single business discipline or function.
6. **Personal.** The skills that require only one person to execute and are not linked to any single business discipline or function.

Even with these definitions, the best-fit classification for a specific skill isn't always obvious. There is a substantial skills overlap between project management and business analysis. I've tried to make an assignment based on relevance and importance to the discipline as the final arbiter.

TABLE A.1 PM Skill/Proficiency Matrix

Sector—PM	Team Member	Task Leader	Associate Manager	Manager	Sr Manager	Consultant	Sr Consultant	Director
Project Management								
Acceptance Testing		2	3	3	4	5	6	4
Agile Project Management		1	2	3	5	6	6	4
Change Management		1	2	3	4	5	6	4
Charter Development		1	3	3	4	5	6	4
Complexity Assessment		2	2	3	4	5	6	6
Cost Estimating		1	2	3	4	6	6	5
Cost Management		1	2	3	4	6	6	5
Critical Path Management		2	3	3	4	5	6	4
Estimation		3	4	4	5	6	6	5
Extreme Project Management		1	2	3	5	6	6	4
Lessons Learned		1	3	4	5	6	6	4
Portfolio Management		1	2	3	4	5	6	6
Project Closeout		1	2	3	4	4	5	5
PM Software		1	1	3	4	5	6	5
Project Notebook		1	2	3	4	4	4	4
Project Organization		1	2	3	4	5	6	4
Project Planning		1	2	3	4	5	6	5
Project Selection		1	3	4	5	6	6	6
Progress Assessment		1	2	3	4	5	6	6
Resource Acquisition			2	3	4	4	4	4
Resource Leveling				3	3	4	4	4
Resource Management			3	3	4	5	5	4

TABLE A.1 *(Continued)*

Sector—PM	Team Member	Task Leader	Associate Manager	Manager	Sr Manager	Consultant	Sr Consultant	Director
Risk Management			3	4	5	6	6	4
Schedule Compression		1	2	3	4	5	6	4
Schedule Development		3	3	4	5	5	5	5
Schedule Management		3	3	3	5	5	5	5
Scope Management			3	4	5	6	6	4
Size Estimating		2	3	4	5	6	6	4
WBS		2	3	4	5	6	6	4
Business Analysis								
Business Rules Analysis		1	2	3	4	4	4	4
Client Change Management		1	2	3	3	4	4	4
Data Analysis	1	2	3	4	4	4	4	4
Data Collection	1	2	3	4	4	4	4	4
Data Flow Diagrams		1	2	3	3	4	4	4
Data Modeling		1	2	3	3	5	5	4
Document Analysis		1	2	3	4	5	5	4
Focus Groups		1	2	3	4	5	5	4
Functional Decomposition		1	2	3	4	5	6	4
Implementation Strategies		1	2	3	4	5	6	5
Interface Analysis		1	2	3	4	5	6	5
Interviewing	1	2	3	4	5	5	5	5
Observation	1	2	3	4	5	5	6	6
Performance Metrics		1	2	3	4	5	6	5
Problem Statement		1	2	3	4	5	6	5

Process Design		1	2	3	5	5	5
Process Improvement	1	2	3	4	5	6	6
Process Modeling	1	2	3	3	4	4	5
Prototyping	1	2	3	4	5	6	5
Requirements Elicitation	1	2	3	4	5	6	6
Requirements Management	1	2	3	4	5	6	6
Requirements Workshop	2	3	4	5	5	6	4
Solution Management	1	2	3	4	5	6	6
Stakeholder Management	1	2	3	4	5	6	6
Structured Walkthrough	1	2	3	4	5	6	6
Survey Design	1	2	3	4	5	6	4
User Stories	1	2	3	4	5	5	5
Vendor Assessment	2	3	4	4	5	6	6
General Management							
Delegation	1	3	4	5	5	5	5
Leadership	1	3	4	5	6	6	6
Managing Change	1	3	4	5	5	5	5
Meeting Management	1	3	4	4	4	4	6
Performance Evaluation	3	4	5	5	6	6	5
Performance Management	2	3	4	5	6	6	6
Quality Management	2	3	4	5	5	5	5
Staff Development	1	2	3	4	5	5	6
Staffing and Hiring	1	2	3	4	4	4	5
Business Functions							
Benchmarking	1	2	3	4	5	6	5
Breakeven Analysis		2	3	4	5	6	6
Budgeting	2	3	4	4	5	5	5

TABLE A.1 *(Continued)*

Sector—PM	Team Member	Task Leader	Associate Manager	Manager	Sr Manager	Consultant	Sr Consultant	Director
Business Case Justification			3	4	5	6	6	6
Business Concepts	1	2	3	4	5	6	6	6
Business Practices		1	2	3	4	5	6	6
Business Strategies		1	2	3	4	5	6	6
Company Products/Services	1	2	3	4	4	5	6	6
Core Systems	1	2	3	4	4	5	6	6
Cost Benefit Analysis		1	2	3	4	5	6	6
Customer Service		1	2	3	4	5	6	6
Feasibility Analysis		1	2	3	4	5	6	6
Financial Analysis		1	2	3	4	5	6	6
Implementation		2	3	4	5	6	6	4
Industry Knowledge		1	2	3	4	5	6	6
Internal Rate of Return		1	2	3	4	6	6	6
Key Performance Indicators		1	2	3	4	5	6	6
Organizational Knowledge	1	2	3	4	5	6	6	6
Planning: Strategic/Tactical		1	2	3	4	5	6	6
Portfolio Management		1	2	3	4	5	6	6
Product/Vendor Evaluation		1	2	3	4	5	6	6
Procedures and Policies	1	2	3	4	5	6	6	6
Return on Investment		1	2	3	4	5	6	6
Software Apps		1	2	3	4	5	6	4
SWOT Analysis		1	2	3	4	5	6	6
Systems Integration		1	2	3	4	6	6	4

Systems Thinking		1	2	3	4	6	6	4
Vision, Mission, Objectives		1	2	3	4	5	6	6
Interpersonal								
Conflict Management	1	2	3	4	5	4	4	6
Flexibility	1	2	3	4	5	5	5	5
Influencing		1	2	3	4	6	6	6
Interpersonal Relationships	1	2	3	4	5	5	5	6
Leadership		1	2	3	4	5	6	6
Managing Priorities	2	3	4	5	5	6	6	6
Negotiating	1	2	3	4	5	5	6	6
Relationship Management		1	2	3	4	5	6	6
Team Building		1	2	3	4	5	5	5
Personal								
Brainstorming	1	2	3	4	5	5	5	5
Creativity	2	3	3	4	4	6	6	6
Decision Making	1	2	3	4	4	6	6	6
Ethics		1	2	3	4	5	5	5
Organization		1	2	3	4	5	5	6
Presentations	1	2	3	4	5	5	6	5
Problem Solving	1	2	3	4	5	5	6	4
Research Methods		1	2	3	4	6	6	4
Root Cause Analysis		1	2	3	4	6	6	4
Teaching	1	2	3	4	5	6	6	6
Trustworthiness	2	3	4	5	5	5	5	6
Verbal Communication	2	3	4	5	5	5	6	6
Written Communication	2	3	4	5	5	5	6	6

TABLE A.2 PM/ba Skill/Proficiency Matrix

Sector—PM/ba	Team Member	Task Leader	Associate Manager	Manager	Sr Manager	Consultant	Sr Consultant	Director
Project Management								
Acceptance Testing		2	3	3	4	5	6	4
Agile Project Management		1	2	3	5	6	6	4
Change Management		1	2	3	4	5	6	4
Charter Development		1	3	3	4	5	6	4
Complexity Assessment		2	2	3	4	5	6	6
Cost Estimating		1	2	3	4	6	6	5
Cost Management		1	2	3	4	6	6	5
Critical Path Management		2	3	3	4	5	6	4
Estimation		3	4	4	5	6	6	5
Extreme Project Management		1	2	3	5	6	6	4
Lessons Learned		1	3	4	5	6	6	4
Portfolio Management		1	2	3	4	5	6	6
Project Closeout		1	2	3	4	4	5	5
PM Software		1	1	3	4	5	6	5
Project Notebook		1	2	3	4	4	4	4
Project Organization		1	2	3	4	5	6	4
Project Planning		1	2	3	4	5	6	5
Project Selection		1	3	4	5	6	6	6
Progress Assessment		1	2	3	4	5	6	6
Resource Acquisition			2	3	4	4	4	4
Resource Leveling				3	3	4	4	4
Resource Management			3	3	4	5	5	4

Risk Management			3	4	5	6	6	4
Schedule Compression		1	2	3	4	5	6	4
Schedule Development		3	3	4	5	5	5	5
Schedule Management		3	3	3	5	5	5	5
Scope Management			3	4	5	6	6	4
Size Estimating		2	3	4	5	6	6	4
WBS		2	3	4	5	6	6	4
Business Analysis								
Business Rules Analysis		2	3	4	4	5	6	4
Client Change Management		2	3	4	5	5	5	5
Data Analysis	1	2	3	4	5	6	6	4
Data Collection	1	2	3	4	5	5	5	4
Data Flow Diagrams	1	2	3	4	5	6	6	4
Data Modeling	1	2	3	4	5	6	6	5
Document Analysis	1	2	3	4	5	6	6	4
Focus Groups		1	2	3	4	5	5	4
Functional Decomposition		1	2	3	4	5	6	4
Implementation Strategies		1	2	3	4	5	6	5
Interface Analysis	1	2	3	4	5	6	6	5
Interviewing	2	3	4	5	5	5	5	5
Observation	2	3	4	5	5	5	6	6
Performance Metrics	1	2	3	4	5	6	6	6
Problem Statement	1	2	3	4	5	6	6	6
Process Design	1	2	3	4	5	6	6	6
Process Improvement	1	2	3	4	5	6	6	6
Process Modeling	1	2	3	4	5	6	6	5
Prototyping	1	2	3	4	5	6	6	6

TABLE A.2 *(Continued)*

Sector—PM/ba	Team Member	Task Leader	Associate Manager	Manager	Sr Manager	Consultant	Sr Consultant	Director
Requirements Elicitation	1	2	3	4	5	5	6	6
Requirements Management	1	2	3	4	5	5	6	6
Requirements Workshop	2	3	4	5	5	6	6	5
Solution Management	1	2	3	4	5	6	6	6
Stakeholder Management	1	2	3	4	5	5	6	6
Structured Walkthrough	1	2	3	4	5	6	6	6
Survey Design	1	2	3	4	5	5	6	4
User Stories	1	2	3	4	5	5	5	5
Vendor Assessment		2	3	4	4	5	6	6
General Management								
Delegation		1	3	4	5	5	5	5
Leadership		1	3	4	5	6	6	6
Managing Change	2	3	4	5	5	5	5	5
Meeting Management	1	2	3	4	5	6	6	6
Performance Evaluation		3	4	5	5	6	6	5
Performance Management		2	3	4	5	6	6	6
Quality Management		2	3	4	5	5	5	5
Staff Development		1	2	3	4	5	5	6
Staffing and Hiring		1	2	3	4	4	4	5
Business Functions								
Benchmarking		1	2	3	4	5	6	5
Breakeven Analysis	1	2	3	4	5	6	6	6
Budgeting		2	3	4	4	5	5	5
Business Case Justification	2	3	4	5	5	6	6	6

Business Concepts	1	2	3	4	5	6	6	6
Business Practices		1	2	3	4	5	6	6
Business Strategies		1	2	3	4	5	6	6
Company Products/Services	1	2	3	4	4	5	6	6
Core Systems	1	2	3	4	4	5	6	6
Cost Benefit Analysis	1	2	3	4	4	5	6	6
Customer Service		1	2	3	4	5	6	6
Feasibility Analysis	1	2	3	4	5	6	6	6
Financial Analysis		1	2	3	4	5	6	6
Implementation	2	3	4	5	5	6	6	6
Industry Knowledge		1	2	3	4	5	6	6
Internal Rate of Return		1	2	3	4	6	6	6
Key Performance Indicator		1	2	3	4	5	6	6
Organizational Knowledge	1	2	3	4	5	6	6	6
Planning: Strategic/Tactical		1	2	3	4	5	6	6
Portfolio Management		1	2	3	4	5	6	6
Product/Vendor Evaluation		1	2	3	4	5	6	6
Procedures and Policies	1	2	3	4	5	6	6	6
Return on Investment	1	2	3	4	5	6	6	6
Software Apps		1	2	3	4	5	6	4
SWOT Analysis		1	2	3	4	5	6	6
Systems Integration	1	2	3	4	5	6	6	4
Systems Thinking	1	2	3	4	5	6	6	4
Vision, Mission, Objectives		1	2	3	4	5	6	6
Interpersonal								
Conflict Management	1	2	3	4	5	4	4	6
Flexibility	1	2	3	4	5	5	5	5
Influencing		1	2	3	4	6	6	6

TABLE A.2 *(Continued)*

Sector—PM/ba	Team Member	Task Leader	Associate Manager	Manager	Sr Manager	Consultant	Sr Consultant	Director
Interpersonal Relationships	1	2	3	4	5	5	5	6
Leadership		1	2	3	4	5	6	6
Managing Priorities	2	3	4	5	5	6	6	6
Negotiating	1	2	3	4	5	5	6	6
Relationship Management		1	2	3	4	5	6	6
Team Building		1	2	3	4	5	5	5
Personal								
Brainstorming	1	2	3	4	5	5	5	5
Creativity	2	3	3	4	4	6	6	6
Decision Making	1	2	3	4	4	6	6	6
Ethics		1	2	3	4	5	5	5
Organization		1	2	3	4	5	5	6
Presentations	1	2	3	4	5	5	6	5
Problem Solving	1	2	3	4	5	5	6	4
Research Methods		1	2	3	4	6	6	4
Root Cause Analysis		1	2	3	4	6	6	4
Teaching	1	2	3	4	5	6	6	6
Trustworthiness	2	3	4	5	5	5	5	6
Verbal Communication	2	3	4	5	5	5	6	6
Written Communication	2	3	4	5	5	5	6	6

TABLE A.3 PM/BA Skill/Proficiency Matrix

Sector—PM/BA	Team Member	Task Leader	Associate Manager	Manager	Sr Manager	Consultant	Sr Consultant	Director
Project Management								
Acceptance Testing		2	3	3	4	5	6	4
Agile Project Management		1	2	3	5	6	6	4
Change Management		1	2	3	4	5	6	4
Charter Development		1	3	3	4	5	6	4
Complexity Assessment		2	2	3	4	5	6	6
Cost Estimating		1	2	3	4	6	6	5
Cost Management		1	2	3	4	6	6	5
Critical Path Management		2	3	3	4	5	6	4
Estimation		3	4	4	5	6	6	5
Extreme Project Management		1	2	3	5	6	6	4
Lessons Learned		1	3	4	5	6	6	4
Portfolio Management		1	2	3	4	5	6	6
Project Closeout		1	2	3	4	4	5	5
PM Software		1	1	3	4	5	6	5
Project Notebook		1	2	3	4	4	4	4
Project Organization		1	2	3	4	5	6	4
Project Planning		1	2	3	4	5	6	5
Project Selection		1	3	4	5	6	6	6
Progress Assessment		1	2	3	4	5	6	6
Resource Acquisition			2	3	4	4	4	4
Resource Leveling				3	3	4	4	4
Resource Management			3	3	4	5	5	4

TABLE A.3 *(Continued)*

Sector—PM/BA	Team Member	Task Leader	Associate Manager	Manager	Sr Manager	Consultant	Sr Consultant	Director
Risk Management			3	4	5	6	6	4
Schedule Compression		1	2	3	4	5	6	4
Schedule Development		3	3	4	5	5	5	5
Schedule Management		3	3	3	5	5	5	5
Scope Management			3	4	5	6	6	4
Size Estimating		2	3	4	5	6	6	4
WBS		2	3	4	5	6	6	4
Business Analysis								
Business Rules Analysis		3	4	5	5	6	6	4
Client Change Management		3	4	5	5	5	5	5
Data Analysis	2	3	4	5	5	6	6	4
Data Collection	2	3	4	5	5	5	5	4
Data Flow Diagrams	2	3	4	5	5	6	6	4
Data Modeling	2	3	4	5	5	6	6	5
Document Analysis	2	3	4	5	5	6	6	4
Focus Groups	1	2	3	4	5	5	5	4
Functional Decomposition	1	2	3	4	5	6	6	4
Implementation Strategies	1	2	3	4	5	5	6	5
Interface Analysis	2	3	4	5	5	6	6	5
Interviewing	2	3	4	5	5	5	5	5
Observation	2	3	4	5	5	5	6	6
Performance Metrics	2	3	4	5	5	6	6	6

Problem Statement	2	3	4	5	5	6	6	6
Process Design	2	3	4	5	5	6	6	6
Process Improvement	2	3	4	5	5	6	6	6
Process Modeling	2	3	4	5	5	6	6	5
Prototyping	2	3	4	5	5	6	6	6
Requirements Elicitation	2	3	4	5	5	5	6	6
Requirements Management	2	3	4	5	5	5	6	6
Requirements Workshop	2	3	4	5	5	6	6	5
Solution Management	2	3	4	5	5	6	6	6
Stakeholder Management	2	3	4	5	5	5	6	6
Structured Walkthrough	2	3	4	5	5	6	6	6
Survey Design	2	3	3	4	5	5	6	4
User Stories	2	3	4	4	5	5	5	5
Vendor Assessment	2	3	4	4	4	5	6	6
General Management								
Delegation	2	3	4	5	5	5	5	5
Leadership	2	3	4	5	5	6	6	6
Managing Change	2	3	4	5	5	5	5	5
Meeting Management	1	2	3	4	5	6	6	6
Performance Evaluation		3	4	5	5	6	6	5
Performance Management		2	3	4	5	6	6	6
Quality Management		2	3	4	5	5	5	5
Staff Development		1	2	3	4	5	5	6
Staffing and Hiring	2	3	4	4	4	4	4	5
Business Functions								
Benchmarking		1	2	3	4	5	6	5
Breakeven Analysis	1	2	3	4	5	6	6	6

TABLE A.3 *(Continued)*

Sector—PM/BA	Team Member	Task Leader	Associate Manager	Manager	Sr Manager	Consultant	Sr Consultant	Director
Budgeting		2	3	4	4	5	5	5
Business Case Justification	2	3	4	5	5	6	6	6
Business Concepts	1	2	3	4	5	6	6	6
Business Practices		1	2	3	4	5	6	6
Business Strategies		1	2	3	4	5	6	6
Company Products/Services	1	2	3	4	4	5	6	6
Core Systems	1	2	3	4	4	5	6	6
Cost Benefit Analysis	1	2	3	4	4	5	6	6
Customer Service	1	2	3	4	5	5	6	6
Feasibility Analysis	1	2	3	4	5	6	6	6
Financial Analysis	1	2	3	4	4	5	6	6
Implementation	2	3	4	5	5	6	6	6
Industry Knowledge	1	2	3	4	4	5	6	6
Internal Rate of Return	1	2	3	4	4	6	6	6
Key Performance Indicators	1	2	3	4	4	5	6	6
Organizational Knowledge	1	2	3	4	5	6	6	6
Planning: Strategic/Tactical		1	2	3	4	5	6	6
Portfolio Management		1	2	3	4	5	6	6
Product/Vendor Evaluation	1	2	3	4	4	5	6	6
Procedures and Policies	1	2	3	4	5	6	6	6
Return on Investment	1	2	3	4	5	6	6	6
Software Apps		2	3	4	5	5	6	4
SWOT Analysis		2	3	4	5	5	6	6

Systems Integration	1	3	4	5	5	6	6	4
Systems Thinking	1	3	4	5	5	6	6	4
Vision, Mission, Objectives		2	3	4	4	5	6	6
Interpersonal								
Conflict Management	1	2	3	4	5	4	4	6
Flexibility	1	2	3	4	5	5	5	5
Influencing		2	3	4	5	6	6	6
Interpersonal Relationships	1	2	3	4	5	5	5	6
Leadership		2	3	4	5	5	6	6
Managing Priorities	2	3	4	5	5	6	6	6
Negotiating	1	2	3	4	5	5	6	6
Relationship Management		2	3	4	5	5	6	6
Team Building		2	3	4	5	5	5	5
Personal								
Brainstorming	1	2	3	4	5	5	5	5
Creativity	2	3	3	4	4	6	6	6
Decision Making	1	2	3	4	4	6	6	6
Ethics		1	2	3	4	5	5	5
Organization		2	3	4	4	5	5	6
Presentations	1	2	3	4	5	5	6	5
Problem Solving	1	2	3	4	5	5	6	4
Research Methods		2	3	4	5	6	6	4
Root Cause Analysis		2	3	4	5	6	6	4
Teaching	1	2	3	4	5	6	6	6
Trustworthiness	2	3	4	5	5	5	5	6
Verbal Communications	2	3	4	5	5	5	6	6
Written Communications	2	3	4	5	5	5	6	6

TABLE A.4 BA/PM Skill/Proficiency Matrix

Sector—BA/PM	Team Member	Task Leader	Associate Manager	Manager	Sr Manager	Consultant	Sr Consultant	Director
Project Management								
Acceptance Testing		2	3	3	4	5	6	4
Agile Project Management		1	2	3	5	6	6	4
Change Management		1	2	3	4	5	6	4
Charter Development		1	3	3	4	5	6	4
Complexity Assessment		2	2	3	4	5	6	6
Cost Estimating		1	2	3	4	6	6	5
Cost Management		1	2	3	4	6	6	5
Critical Path Management		2	3	3	4	5	6	4
Estimation		3	4	4	5	6	6	5
Extreme Project Management		1	2	3	5	6	6	4
Lessons Learned		1	3	4	5	6	6	4
Portfolio Management		1	2	3	4	5	6	6
Project Closeout		1	2	3	4	4	5	5
PM Software		1	1	3	4	5	6	5
Project Notebook		1	2	3	4	4	4	4
Project Organization		1	2	3	4	5	6	4
Project Planning		1	2	3	4	5	6	5
Project Selection		1	3	4	5	6	6	6
Progress Assessment		1	2	3	4	5	6	6
Resource Acquisition			2	3	4	4	4	4
Resource Leveling				3	3	4	4	4
Resource Management			3	3	4	5	5	4

Risk Management			3	4	5	6	6	4
Schedule Compression		1	2	3	4	5	6	4
Schedule Development		3	3	4	5	5	5	5
Schedule Management		3	3	3	5	5	5	5
Scope Management			3	4	5	6	6	4
Size Estimating		2	3	4	5	6	6	4
WBS		2	3	4	5	6	6	4
Business Analysis								
Business Rules Anal.		3	4	5	5	6	6	4
Client Change Management		3	4	5	5	5	5	5
Data Analysis	2	3	4	5	5	6	6	4
Data Collection	2	3	4	5	5	5	5	4
Data Flow Diagrams	2	3	4	5	5	6	6	4
Data Modeling	2	3	4	5	5	6	6	5
Document Analysis	2	3	4	5	5	6	6	4
Focus Groups	1	2	3	4	5	5	5	4
Functional Decomposition	1	2	3	4	5	6	6	4
Implementation Strategies	1	2	3	4	5	5	6	5
Interface Analysis	2	3	4	5	5	6	6	5
Interviewing	2	3	4	5	5	5	5	5
Observation	2	3	4	5	5	5	6	6
Performance Metrics	2	3	4	5	5	6	6	6
Problem Statement	2	3	4	5	5	6	6	6
Process Design	2	3	4	5	5	6	6	6
Process Improvement	2	3	4	5	5	6	6	6
Process Modeling	2	3	4	5	5	6	6	5

TABLE A.4 *(Continued)*

Sector—BA/PM	Team Member	Task Leader	Associate Manager	Manager	Sr Manager	Consultant	Sr Consultant	Director
Prototyping	2	3	4	5	5	6	6	6
Requirements Elicitation	2	3	4	5	5	5	6	6
Requirements Management	2	3	4	5	5	5	6	6
Requirements Workshop	2	3	4	5	5	6	6	5
Solution Management	2	3	4	5	5	6	6	6
Stakeholder Management	2	3	4	5	5	5	6	6
Structured Walkthrough	2	3	4	5	5	6	6	6
Survey Design	2	3	3	4	5	5	6	4
User Stories	2	3	4	4	5	5	5	5
Vendor Assessment	2	3	4	4	4	5	6	6
General Management								
Delegation	2	3	4	5	5	5	5	5
Leadership	2	3	4	5	5	6	6	6
Managing Change	2	3	4	5	5	5	5	5
Meeting Management	1	2	3	4	5	6	6	6
Performance Evaluation		3	4	5	5	6	6	5
Performance Management		2	3	4	5	6	6	6
Quality Management		2	3	4	5	5	5	5
Staff Development		1	2	3	4	5	5	6
Staffing and Hiring	2	3	4	4	4	4	4	5
Business Functions								
Benchmarking		1	2	3	4	5	6	5

Breakeven Analysis	1	2	3	4	5	6	6	6
Budgeting		2	3	4	4	5	5	5
Business Case Justification	2	3	4	5	5	6	6	6
Business Concepts	1	2	3	4	5	6	6	6
Business Practices		1	2	3	4	5	6	6
Business Strategies		1	2	3	4	5	6	6
Company Products/Services	1	2	3	4	4	5	6	6
Core Systems	1	2	3	4	4	5	6	6
Cost Benefit Analysis	1	2	3	4	4	5	6	6
Customer Service	1	2	3	4	5	5	6	6
Feasibility Analysis	1	2	3	4	5	6	6	6
Financial Analysis	1	2	3	4	4	5	6	6
Implementation	2	3	4	5	5	6	6	6
Industry Knowledge	1	2	3	4	4	5	6	6
Internal Rate of Return	1	2	3	4	4	6	6	6
Key Performance Indicators	1	2	3	4	4	5	6	6
Organizational Knowledge	1	2	3	4	5	6	6	6
Planning: Strategic/Tactical		1	2	3	4	5	6	6
Portfolio Management		1	2	3	4	5	6	6
Product/Vendor Evaluation	1	2	3	4	4	5	6	6
Procedures and Policies	1	2	3	4	5	6	6	6
Return on Investment	1	2	3	4	5	6	6	6
Software Apps		2	3	4	5	5	6	4
SWOT Analysis		2	3	4	5	5	6	6
Systems Integration	1	3	4	5	5	6	6	4
Systems Thinking	1	3	4	5	5	6	6	4
Vision, Mission, Objectives		2	3	4	4	5	6	6

TABLE A.4 *(Continued)*

Sector—BA/PM	Team Member	Task Leader	Associate Manager	Manager	Sr Manager	Consultant	Sr Consultant	Director
Interpersonal								
Conflict Management	1	2	3	4	5	4	4	6
Flexibility	1	2	3	4	5	5	5	5
Influencing		2	3	4	5	6	6	6
Interpersonal Relationships	1	2	3	4	5	5	5	6
Leadership		2	3	4	5	5	6	6
Managing Priorities	2	3	4	5	5	6	6	6
Negotiating	1	2	3	4	5	5	6	6
Relationship Management		2	3	4	5	5	6	6
Team Building		2	3	4	5	5	5	5
Personal								
Brainstorming	1	2	3	4	5	5	5	5
Creativity	2	3	3	4	4	6	6	6
Decision Making	1	2	3	4	4	6	6	6
Ethics		1	2	3	4	5	5	5
Organization		2	3	4	4	5	5	6
Presentations	1	2	3	4	5	5	6	5
Problem Solving	1	2	3	4	5	5	6	4
Research Methods		2	3	4	5	6	6	4
Root Cause Analysis		2	3	4	5	6	6	4
Teaching	1	2	3	4	5	6	6	6
Trustworthiness	2	3	4	5	5	5	5	6
Verbal Communication	2	3	4	5	5	5	6	6
Written Communication	2	3	4	5	5	5	6	6

TABLE A.5 BA/pm Skill/Proficiency Matrix

Sector—BA/pm	Team Member	Task Leader	Associate Manager	Manager	Sr Manager	Consultant	Sr Consultant	Director
Project Management								
Acceptance Testing		2	3	3	4	5	5	4
Agile Project Management		1	2	3	4	5	6	5
Change Management		1	2	3	4	5	5	4
Charter Development		1	2	3	4	5	5	4
Complexity Assessment		1	2	3	4	5	5	5
Cost Estimating		1	2	3	4	5	5	5
Cost Management		1	2	3	4	6	6	5
Critical Path Management		1	2	3	4	5	5	4
Estimation		2	3	4	4	5	6	5
Extreme Project Management		1	2	3	4	5	6	5
Lessons Learned		2	3	4	4	5	5	4
Portfolio Management			2	3	4	5	5	5
Project Closeout		1	2	3	4	4	5	5
PM Software		1	2	3	4	5	5	5
Project Notebook		1	2	3	4	5	5	5
Project Organization		1	2	3	4	5	5	5
Project Planning		1	2	3	4	5	5	5
Project Selection		1	2	3	4	5	5	5
Progress Assessment		1	2	3	4	5	6	5
Resource Acquisition		1	2	3	4	4	4	4
Resource Leveling		1	2	3	3	4	4	4
Resource Management		1	2	3	3	4	4	4

TABLE A.5 *(Continued)*

Sector—BA/pm	Team Member	Task Leader	Associate Manager	Manager	Sr Manager	Consultant	Sr Consultant	Director
Risk Management			2	3	4	5	6	4
Schedule Compression		1	2	3	4	4	5	4
Schedule Development		1	2	3	4	5	5	5
Schedule Management		1	2	3	4	5	5	5
Scope Management		2	3	4	4	5	6	4
Size Estimating		2	3	4	4	5	5	4
WBS		2	3	4	5	5	5	4
Business Analysis								
Business Rules Analysis		1	2	3	4	4	4	4
Client Change Management		1	2	3	3	4	4	4
Data Analysis	1	2	3	4	4	4	4	4
Data Collection	1	2	3	4	4	4	4	4
Data Flow Diagrams		1	2	3	3	4	4	4
Data Modeling		1	2	3	3	5	5	4
Document Analysis		1	2	3	4	5	5	4
Focus Groups		1	2	3	4	5	5	4
Functional Decomposition		1	2	3	4	5	6	4
Implementation Strategies		1	2	3	4	5	6	5
Interface Analysis		1	2	3	4	5	6	5
Interviewing	1	2	3	4	5	5	5	5
Observation	1	2	3	4	5	5	6	6
Performance Metrics		1	2	3	4	5	6	5
Problem Statement		1	2	3	4	5	6	5

Process Design		1	2	3	5	5	5
Process Improvement	1	2	3	4	5	6	6
Process Modeling	1	2	3	3	4	4	5
Prototyping	1	2	3	4	5	6	5
Requirements Elicitation	1	2	3	4	5	6	6
Requirements Management	1	2	3	4	5	6	6
Requirements Workshop	2	3	4	5	5	6	4
Solution Management	1	2	3	4	5	6	6
Stakeholder Management	1	2	3	4	5	6	6
Structured Walkthrough	1	2	3	4	5	6	6
Survey Design	1	2	3	4	5	6	4
User Stories	1	2	3	4	5	5	5
Vendor Assessment	2	3	4	4	5	6	6
General Management							
Delegation	1	3	4	5	5	5	5
Leadership	1	3	4	5	6	6	6
Managing Change	1	3	4	5	5	5	5
Meeting Management	1	3	4	4	4	4	6
Performance Evaluation	3	4	5	5	6	6	5
Performance Management	2	3	4	5	6	6	6
Quality Management	2	3	4	5	5	5	5
Staff Development	1	2	3	4	5	5	6
Staffing and Hiring	1	2	3	4	4	4	5
Business Functions							
Benchmarking	1	2	3	4	5	6	5
Breakeven Analysis		2	3	4	5	6	6

TABLE A.5 *(Continued)*

Sector—BA/pm	Team Member	Task Leader	Associate Manager	Manager	Sr Manager	Consultant	Sr Consultant	Director
Budgeting		2	3	4	4	5	5	5
Business Case Justification			3	4	5	6	6	6
Business Concepts	1	2	3	4	5	6	6	6
Business Practices		1	2	3	4	5	6	6
Business Strategies		1	2	3	4	5	6	6
Company Products/Services	1	2	3	4	4	5	6	6
Core Systems	1	2	3	4	4	5	6	6
Cost Benefit Analysis		1	2	3	4	5	6	6
Customer Service		1	2	3	4	5	6	6
Feasibility Analysis		1	2	3	4	5	6	6
Financial Analysis		1	2	3	4	5	6	6
Implementation		2	3	4	5	6	6	4
Industry Knowledge		1	2	3	4	5	6	6
Internal Rate of Return		1	2	3	4	6	6	6
Key Performance Indicators		1	2	3	4	5	6	6
Organizational Knowledge	1	2	3	4	5	6	6	6
Planning: Strategic/Tactical		1	2	3	4	5	6	6
Portfolio Management		1	2	3	4	5	6	6
Product/Vendor Evaluation		1	2	3	4	5	6	6
Procedures and Policies	1	2	3	4	5	6	6	6
Return on Investment		1	2	3	4	5	6	6
Software Apps		1	2	3	4	5	6	4
SWOT Analysis		1	2	3	4	5	6	6

Systems Integration		1	2	3	4	6	6	4
Systems Thinking		1	2	3	4	6	6	4
Vision, Mission, Objectives		1	2	3	4	5	6	6
Interpersonal								
Conflict Management	1	2	3	4	5	4	4	6
Flexibility	1	2	3	4	5	5	5	5
Influencing		1	2	3	4	6	6	6
Interpersonal Relationship	1	2	3	4	5	5	5	6
Leadership		1	2	3	4	5	6	6
Managing Priorities	2	3	4	5	5	6	6	6
Negotiating	1	2	3	4	5	5	6	6
Relationship Management		1	2	3	4	5	6	6
Team Building		1	2	3	4	5	5	5
Personal								
Brainstorming	1	2	3	4	5	5	5	5
Creativity	2	3	3	4	4	6	6	6
Decision Making	1	2	3	4	4	6	6	6
Ethics		1	2	3	4	5	5	5
Organization		1	2	3	4	5	5	6
Presentations	1	2	3	4	5	5	6	5
Problem Solving	1	2	3	4	5	5	6	4
Research Methods		1	2	3	4	6	6	4
Root Cause Analysis		1	2	3	4	6	6	4
Teaching	1	2	3	4	5	6	6	6
Trustworthiness	2	3	4	5	5	5	5	6
Verbal Communication	2	3	4	5	5	5	6	6
Written Communication	2	3	4	5	5	5	6	6

TABLE A.6 BA Skill/Proficiency Matrix

Sector—BA	Team Member	Task Leader	Associate Manager	Manager	Sr Manager	Consultant	Sr Consultant	Director
Project Management								
Acceptance Testing		1	2	3	3	4	4	4
Agile Project Management			1	2	3	5	5	5
Change Management		1	2	3	4	5	5	4
Charter Development		1	2	3	3	4	4	4
Complexity Assessment			1	2	3	5	5	5
Cost Estimating			1	2	3	4	5	5
Cost Management			1	2	3	4	5	5
Critical Path Management			1	2	3	3	5	4
Estimation		1	2	3	4	5	5	5
Extreme Project Management			1	2	3	5	5	5
Lessons Learned		1	2	3	4	4	4	4
Portfolio Management			1	2	3	4	4	4
Project Closeout			1	2	3	4	4	4
PM Software			1	3	4	4	4	4
Project Notebook			1	2	3	4	4	4
Project Organization			1	2	3	4	4	4
Project Planning			1	2	3	3	4	4
Project Selection			1	2	3	4	5	5
Progress Assessment		1	2	3	4	5	6	6
Resource Acquisition			1	2	3	4	4	4
Resource Leveling			1	2	3	4	4	4
Resource Management			1	2	3	4	4	4

Risk Management			2	3	4	5	6	4
Schedule Compression			1	2	3	3	4	4
Schedule Development		1	2	3	4	5	5	5
Schedule Management		1	2	3	4	5	5	5
Scope Management			2	3	4	5	5	4
Size Estimating		1	2	3	3	5	5	4
WBS		1	2	3	4	5	5	4
Business Analysis								
Business Rules Analysis		1	2	3	4	4	4	4
Client Change Management		1	2	3	3	4	4	4
Data Analysis	1	2	3	4	4	4	4	4
Data Collection	1	2	3	4	4	4	4	4
Data Flow Diagrams		1	2	3	3	4	4	4
Data Modeling		1	2	3	3	5	5	4
Document Analysis		1	2	3	4	5	5	4
Focus Groups		1	2	3	4	5	5	4
Functional Decomposition		1	2	3	4	5	6	4
Implementation Strategies		1	2	3	4	5	6	5
Interface Analysis		1	2	3	4	5	6	5
Interviewing	1	2	3	4	5	5	5	5
Observation	1	2	3	4	5	5	6	6
Performance Metrics		1	2	3	4	5	6	5
Problem Statement		1	2	3	4	5	6	5
Process Design			1	2	3	5	5	5
Process Improvement		1	2	3	4	5	6	6
Process Modeling		1	2	3	3	4	4	5
Prototyping		1	2	3	4	5	6	5

TABLE A.6 *(Continued)*

Sector—BA	Team Member	Task Leader	Associate Manager	Manager	Sr Manager	Consultant	Sr Consultant	Director
Requirements Elicitation		1	2	3	4	5	6	6
Requirements Management		1	2	3	4	5	6	6
Requirements Workshop		2	3	4	5	5	6	4
Solution Management		1	2	3	4	5	6	6
Stakeholder Management		1	2	3	4	5	6	6
Structured Walkthrough		1	2	3	4	5	6	6
Survey Design		1	2	3	4	5	6	4
User Stories		1	2	3	4	5	5	5
Vendor Assessment		2	3	4	4	5	6	6
General Management								
Delegation		1	3	4	5	5	5	5
Leadership		1	3	4	5	6	6	6
Managing Change		1	3	4	5	5	5	5
Meeting Management		1	3	4	4	4	4	6
Performance Evaluation		3	4	5	5	6	6	5
Performance Management		2	3	4	5	6	6	6
Quality Management		2	3	4	5	5	5	5
Staff Development		1	2	3	4	5	5	6
Staffing and Hiring		1	2	3	4	4	4	5
Business Functions								
Benchmarking		1	2	3	4	5	6	5
Breakeven Analysis			2	3	4	5	6	6

Budgeting		2	3	4	4	5	5	5
Business Case Justification			3	4	5	6	6	6
Business Concepts	1	2	3	4	5	6	6	6
Business Practices		1	2	3	4	5	6	6
Business Strategies		1	2	3	4	5	6	6
Company Products/Services	1	2	3	4	4	5	6	6
Core Systems	1	2	3	4	4	5	6	6
Cost Benefit Analysis		1	2	3	4	5	6	6
Customer Service		1	2	3	4	5	6	6
Feasibility Analysis		1	2	3	4	5	6	6
Financial Analysis		1	2	3	4	5	6	6
Implementation		2	3	4	5	6	6	4
Industry Knowledge		1	2	3	4	5	6	6
Internal Rate of Return		1	2	3	4	6	6	6
Key Performance Indicators		1	2	3	4	5	6	6
Organizational Knowledge	1	2	3	4	5	6	6	6
Planning: Strategic/Tactical		1	2	3	4	5	6	6
Portfolio Management		1	2	3	4	5	6	6
Product/Vendor Evaluation		1	2	3	4	5	6	6
Procedures and Policies	1	2	3	4	5	6	6	6
Return on Investment		1	2	3	4	5	6	6
Software Apps		1	2	3	4	5	6	4
SWOT Analysis		1	2	3	4	5	6	6
Systems Integration		1	2	3	4	6	6	4
Systems Thinking		1	2	3	4	6	6	4
Vision, Mission, Objectives		1	2	3	4	5	6	6

TABLE A.6 *(Continued)*

Sector—BA	Team Member	Task Leader	Associate Manager	Manager	Sr Manager	Consultant	Sr Consultant	Director
Interpersonal								
Conflict Management	1	2	3	4	5	4	4	6
Flexibility	1	2	3	4	5	5	5	5
Influencing		1	2	3	4	6	6	6
Interpersonal Relationships	1	2	3	4	5	5	5	6
Leadership		1	2	3	4	5	6	6
Managing Priorities	2	3	4	5	5	6	6	6
Negotiating	1	2	3	4	5	5	6	6
Relationship Management		1	2	3	4	5	6	6
Team Building		1	2	3	4	5	5	5
Personal								
Brainstorming	1	2	3	4	5	5	5	5
Creativity	2	3	3	4	4	6	6	6
Decision Making	1	2	3	4	4	6	6	6
Ethics		1	2	3	4	5	5	5
Organization		1	2	3	4	5	5	6
Presentations	1	2	3	4	5	5	6	5
Problem Solving	1	2	3	4	5	5	6	4
Research Methods		1	2	3	4	6	6	4
Root Cause Analysis		1	2	3	4	6	6	4
Teaching	1	2	3	4	5	6	6	6
Trustworthiness	2	3	4	5	5	5	5	6
Verbal Communication	2	3	4	5	5	5	6	6
Written Communication	2	3	4	5	5	5	6	6

APPENDIX B

PM and BA Training Provider Courses

I have searched the Internet looking for training providers who offer a significant portfolio of instructor-led courses in project management and business analysis. My purpose is to find any providers who are positioned to offer courses that reflect the PM/BA professionals defined in this book. As expected, I didn't expect to find much—at least not at this time. In Appendix C I offer a few possibilities for such programs and courses.

Table B.1 offers a partial list of training providers who have significant offerings in PM and/or BA. The list does not include every trainer in the PM or BA space. The vendors whose offerings are listed are coded in this way:

AMA	American Management Association
BS	Babbage Simmel
B2T	B2T Training
BUT	Boston University Training
ESI	ESI International
GKN	Global Knowledge Network
LTI	Learning Tree International
LQ	LearnQuest
MC	Management Concepts
PMC	PM Centers USA
PRG	Pierson Requirements Group
WL	Watermark Learning

The numbers in the cells denote the number of instructor-led days. If the cell is blank, I did not find any relevant instructor-led course. An "X" means the vendor has a program but its duration was not known at the time the data was collected. In a survey like this, there is always the problem of course titles and course content. I made several judgment calls and tried to present the information as accurately as possible.

TABLE B.1 Training Providers PM and BA Courses

Course Name	AMA	BS	B2T	BUT	ESI	GKN	LTI	LQ	MC	PMC	PRG	WL
PM/BA												
BA												
Building an Effective Business Case	X						3		2			
Building a Successful Business Analysis Work Plan			3				3					
Business Analysis Boot Camp										4		
Business Process Analysis, Innovation, and Design	3			X	3		3	3		2		
Business Process Management				X								
Business Process Reengineering							4				2	
Business Requirements Using Joint Application Design (JAD)											4	
CBAP Exam Preparation			4		2		4			4		3
Consulting Skills to Solve Business Problems												2
Data Modeling					3			2			2	2
Data Modeling—Advanced												2
Defining Business Systems with UML										X	2	
Designing and Facilitating JAD Workshops										2		
Developing Requirements and Selecting Software Package Solutions											2	
Developing Requirements with Use Cases				X					1	2		2
Eliciting Business Requirements				X								2
Facilitation Techniques for Requirements Development			3		2			4				2
Foundations of Business Analysis		X	4	X	3				3			3

How to Gather and Document User Requirements			3	X	4		4					
Influencing without Authority				X								2
Managing the Outsourced Organization		X										
Process Modeling Management	3		4		3		4	3			2	2
Requirements Management Workshop	X			X					3	2		3
Requirements Risk										2		
Requirements Validation			2						3	2		
Strategic Enterprise Analysis					3		3					
Testing Techniques for Tracing and Validating Requirements					3			3		2		2
Use Case Modeling				X	4			3				2
PM												
Acceptance Testing											2	
Advanced IT Project Management				X								
Advanced Project Management									3	2		
Advanced Risk Management				X								
Agile Project Management					2					2		
Agile Projects Using Scrum				X								
Agile Requirements										2	4	
Aligning Project Management with Organizational Strategy				X	3		4					
Building a Successful PMO				X			3					
Capital Asset Planning									3			
CAPM Exam Preparation							3					
Change Management				X								

TABLE B.1 *(Continued)*

Course Name	AMA	BS	B2T	BUT	ESI	GKN	LTI	LQ	MC	PMC	PRG	WL
Conducting Effective Collaborative Meetings											2	
Contract Management					3		3					
Cost Estimating									2			
Earned Value				X	2				2			
Human Resources Management				X				2				
Intelligent Disobedience: Difference between Good and GREAT Project Managers												2
IT Risk					3							
Leading Complex Projects					3		2		2			
Leading High-Performance Project Teams				X	2		4					
Leading Project Managers	X			X	2							
Lean-Agile Release Planning								2				
Lean Software Development								2				
Lean Software Development for Management								1				
Managing Global Projects					3							
Managing Multiple Projects	2								3			
Managing Outsource and Offshore Projects										2		
Managing Projects	3				3		4		3		2	3
Managing IT Projects	3				3			2	4	2		3
Managing Small Projects												2
Microsoft Project				X	X		3		1	2		2
Microsoft Server					X		3					
Negotiation Skills					3							

PgMP Exam Preparation							3					
PMP Certification	3				2		5		5	5		3
Prince2				X			5					5
Procurement and Cost Management				X				2	3			
Program Management				X	3							
Project Budgeting and Estimating							3					
Project Leadership	3			X	3				4			
Project Management Applications					4				1	3		
Project Management for Non–Project Managers	2											
Project Management Simulation				X					5			
Project Management Workshop	5											
Project Planning and Estimating											2	2
Project Sponsor Workshop												1
Project Time Management				X								
Quality				X	3			2	3			
Requirements Management	2				3				3			
Risk				X	3		4	2	3			1
Scheduling and Cost Control					4		3	2	3			
Scope Management				X								
SharePoint							3					
Software Testing					3							
Systems Integration					3							
Technical Project Management	3											
Troubled Projects				X	3							2
User Acceptance Testing											2	
Writing Statements of Work					3							

APPENDIX C

PM/BA Curriculum

This innovative curriculum reflects the integration of several disciplines, among which I would include:

- Strategic planning
- Enterprise architecture
- Business analysis
- Information technology management
- Business process management
- Portfolio management
- Project management

These seven disciplines are related to one another as shown in Figure C.1.

If the curriculum is simply a collection of courses from each discipline with no integration or integrating courses, it will not do the job. In a certain sense, PM/BA and BA/PM professionals are generalists with application knowledge across project management and business analysis and perhaps some of the other business disciplines shown in Figure C.1. These professionals work at the operational, tactical, and strategic levels, and the curriculum must reflect that scope. That immediately raises questions about the organizational structure of the curriculum. First of all, should it be discipline-centric or enterprise-centric? I've had experience designing curricula that integrate several disciplines and settled on the enterprise-centric model that is team taught. Some courses in this team-taught model are introductions to a discipline and can be taught by a faculty who are more specialists in one of the seven disciplines than generalists across multiple disciplines. Other courses will be taught by a team of faculty members. This requires collaboration and planning, but I have been integrally involved in such efforts and they can be very effective. However, every faculty team

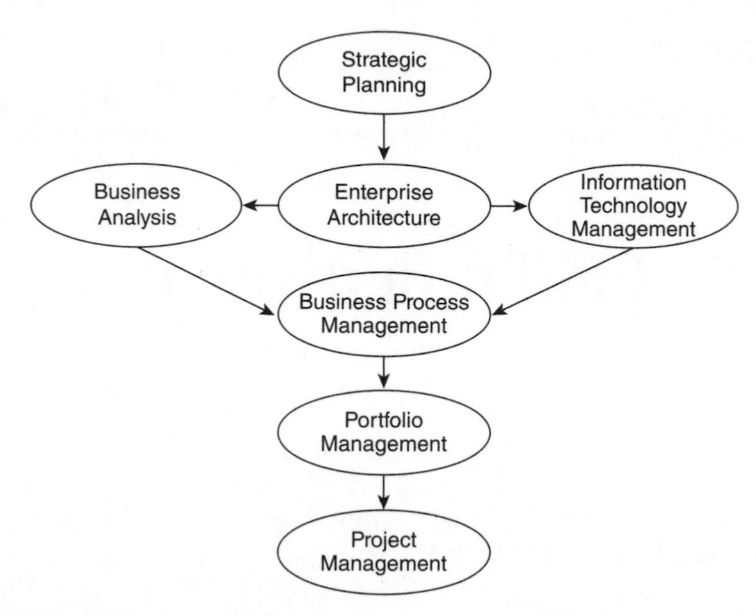

FIGURE C.1 Discipline Integration

member must understand how his or her discipline integrates across the enterprise model and teach from that perspective.

Each course is designed around a learning model that is team driven. When on the job, most assignments will be completed using some form of team structure so it is only appropriate that students learn in a team structure. Adopting this approach exposes students to the team environment and how to deal with the issues that arise, and do so in a nonthreatening environment.

At the course level, the major courses might be:

- Introduction to Contemporary Organizations
- Managing Data and Information across the Organization
- Business Process Management
- Creative Models for Solving Business Problems
- Introduction to Project and Portfolio Management
- An Enterprise-Wide Project

These six courses can form the required core courses of a major at the undergraduate or graduate level or an advanced graduate-level certificate. The course descriptions including their major topic list are shown in the next sections.

Introduction to Contemporary Organizations

Course Description

This is the introductory course in the program. It examines the organization from strategic, tactical, and operational perspectives using a model that covers strategic planning, enterprise architecture, business analysis, information technology, business process management, portfolio management, and project management. The purpose of the course is to help the student understand how the PM/BA interacts with the organization from the perspective of the business processes that define it.

Prerequisite(s)

None

Major Topics

- A process model of the contemporary organization
- The fundamentals, process, and structure of strategic planning
- Strengths, Weaknesses, Opportunities, Threats (SWOT) analysis as input to the strategic planning process
- How the strategic plan influences the organization
- Analyzing the alignment of the business to the strategic plan
- Adjusting the IT plan to support the strategic plan
- Identifying opportunities to design, develop, and improve business processes
- Establishing a project portfolio management process to support the strategic plan
- Proposing and managing projects to maximize their business value

Managing Data and Information across the Organization

Course Description

This course examines data and information as a corporate asset and presents a process model that integrates business analysis, enterprise architecture, and information technology at the management level. The course takes a strategic-, tactical-, and operational-level view.

Prerequisite(s)

Introduction to the Contemporary Organization

Major Topics

- Process model for linking business analysis, enterprise architecture, and information technology
- Aligning business processes and architecture with the strategic plan
- The role of information in driving the strategic plan
- Fundamentals of business analysis
- Enterprise and business process architecture
- Information technology as the enabler

Business Process Management

Course Description

Business Process Management (BPM) focuses on modeling, managing, and optimizing business processes. In order to do this, the disciplines of systems design and development, business analysis, and project management must integrate into a single discipline. In effect, this creates a process-centric layer that can effectively respond to changing business needs. Participants in this introductory course will learn several approaches to business modeling and process management.

Prerequisite(s)

Business Analysis
Information Technology Management

Major Topics

- Enterprise-level view of business process management
- Linking business processes to strategic plans
- Analyzing and scoping a business process problem
- Modeling the "As Is" and "To Be" business process
- Business process modeling and management tools
- Establishing quantitative performance measures
- Validating and analyzing a business process
- Business process improvement projects and portfolios
- Implementing business process change

Introduction to Project and Portfolio Management

Course Description

This is an introductory course in project and portfolio management for BAs. Contemporary projects are often projects of high complexity and uncertainty. To be an effective BA on such projects requires at least a working knowledge of project and portfolio management. The person might be the PM on such projects; at the least, he or she will have a co-PM role and responsibility on the project.

Prerequisite(s)

Business Process Management

Major Topics

- Project Landscape
- Project Management Life Cycle Models (PMLC)
  - Traditional Project Management
  - Agile Project Management
  - Extreme Project Management
- The BA Role and Responsibility in the PMLC
- PM and BA collaboration
- Portfolio Management Processes
  - Traditional Model
  - Agile Model

Creative Models for Solving Business Problems

Course Description

The BA will be involved in and perhaps manage business process improvement projects. In many cases the improvement of a business process has eluded discovery, and a project has been commissioned to discover ways to improve the performance of a business process in order to add business value. To be successful in such projects requires a solid working knowledge of problem solving.

Prerequisite(s)

Business Process Management
Project and Portfolio Management

Major Topics

- Definition of a Business Problem
- Comparative Analysis of Problem-Solving Process Models
- Graphical Tools for Problem Solving

An Enterprise-Wide Project

Course Description

This is the capstone course in the program. A student team of four to six members will work with the senior management team of a local organization. The purpose of the project is to assess the degree to which the enterprise model has been implemented. The assessment will include a documented description of the seven processes as they have been implemented in the organization. An analysis of the strengths, weaknesses, opportunities, and threats will be included. The project will span two semesters and for credit purposes will be the equivalent of two courses.

Prerequisite(s)

Completion of all other courses or concurrent registration in the last course(s)

Major Topics

- Assessment and documentation of the implementation of the seven disciplines
- SWOT analysis
- Report preparation and presentation

About the Author

Robert K. Wysocki, Ph.D., has over 45 years' experience as a project management consultant and trainer, information systems manager, systems and management consultant, author, training developer, and provider. He has written 18 books on project management and information systems management. One of his books, *Effective Project Management: Traditional,*

Agile, Extreme, 5th ed. (Hoboken, NJ: John Wiley & Sons, 2009), has been a bestseller and is recommended by the Project Management Institute for the library of every project manager. He has over 30 publications in professional and trade journals and has made more than 100 presentations at professional and trade conferences and meetings. He has developed more than 20 project management courses and trained over 10,000 project managers.

From 1963 to 1970 he was a systems consultant for one of the world's largest electronics components manufacturers. In that capacity he designed and implemented several computer-based manufacturing and quality control systems. From 1970 to 1990 he held a number of positions in both state-supported and private institutions in higher education as MBA Director, Associate Dean of Business, Dean of Computers and Information Systems, Director of Academic Computing, CIO, and Senior Planner.

In 1990 he founded Enterprise Information Insights, Inc. (EII), a project management consulting and training practice specializing in project management methodology design and integration, project support office establishment, the development of training curriculum, and the development of a portfolio of assessment tools focused on organizations, project teams, and individuals. His client list includes AT&T, Aetna, Babbage Simmel, BMW, British Computer Society, Boston University Corporate Education Center, Computerworld, Converse Shoes, Czechoslovakian government, Data General, Digital, Eli Lilly, Harvard Community Health Plan, IBM, J. Walter Thompson, Ohio State University, Peoples Bank, Sapient Corporation, The Limited, The State of Ohio, Travelers Insurance, TVA, University of California–Santa Cruz, U.S. Coast Guard Academy, Wal-Mart, Wells Fargo, ZTE, and several others.

He is a Senior Consultant at the Cutter Consortium, where he is an active member of the Agile Project Management Practice. He is the Series Editor of Effective Project Management for Artech House, a publisher to the technical and engineering professions. He was a founding member of the Agile Project Leadership Network and served as its first Vice President and President Elect, a member of asapm, the Agile Alliance, the Project Summit Business Analyst World Executive Advisory Board, the Project Management Institute, the American Society of Training & Development, and the Society of Human Resource Management. He is past Association Vice President of AITP (formerly DPMA). He earned a BA in Mathematics from the University of Dallas and an MS and PhD in Mathematical Statistics from Southern Methodist University.

Index